Durability Design of Concrete Structures in Severe Environments

Second Edition

Durability Design of Concrete Structures in Severe Environments

Second Edition

Odd E. Gjørv

CRC Press
Taylor & Francis Group
Boca Raton London New York

CRC Press is an imprint of the
Taylor & Francis Group, an **informa** business
A SPON BOOK

CRC Press
Taylor & Francis Group
6000 Broken Sound Parkway NW, Suite 300
Boca Raton, FL 33487-2742

First issued in paperback 2017

© 2014 by Taylor & Francis Group, LLC
CRC Press is an imprint of Taylor & Francis Group, an Informa business

No claim to original U.S. Government works
Version Date: 20131212

ISBN 13: 978-1-138-07390-6 (pbk)
ISBN 13: 978-1-4665-8729-8 (hbk)

Visit the Taylor & Francis Web site at
http://www.taylorandfrancis.com

and the CRC Press Web site at
http://www.crcpress.com

Contents

5 Additional strategies and protective measures   131

6 Concrete quality control and quality assurance   153

7 Achieved construction quality   169

Preface

Concrete structures in severe environments include a variety of structures in various types of environments. Although several deteriorating processes such as alkali–aggregate reactions, freezing and thawing, and chemical attack still represent severe challenges and problems to many important concrete structures, rapid development in concrete technology in recent years has made it easier to control such deteriorating processes. Also, for new concrete structures in severe environments, the applied concrete is normally so dense that concrete carbonation does not represent any practical problem. For concrete structures in chloride-containing environments, however, chloride ingress and premature corrosion of embedded steel still appear to be a most difficult and severe challenge to the durability and performance of many important concrete infrastructures. In recent years, there has also been a rapid increase in the use of de-icing salt and rapid development on concrete structures in marine environments.

In order to obtain increased and better control of chloride ingress and corrosion of embedded steel, improved procedures and specifications for proper combinations of concrete quality and concrete cover are very important. Upon completion of new concrete structures, however, the achieved construction quality typically shows high scatter and variability, and, in severe environments, any weaknesses and deficiencies will soon be revealed, whatever durability specifications and materials have been applied. Therefore, improved procedures for quality control and quality assurance during concrete construction are also very important.

To a certain extent, a probability approach to the durability design can accommodate the high scatter and variability. However, a numerical approach alone is not sufficient to ensure the durability. In order to obtain a more controlled and improved durability, it is also essential to specify performance-based durability requirements that can be verified and controlled for proper quality assurance during concrete construction. Documentation of achieved construction quality and compliance with the specified durability should be the keys to any rational approach to more controlled and increased durability and service life of concrete structures in severe environments. Better procedures for condition assessment and preventive maintenance should also be essential, and such procedures should help provide the ultimate basis for achieving more controlled durability and service life of concrete structures.

In recent years, an increased number of owners of concrete structures have realized that even small additional costs, in order to obtain an increased and more controlled durability beyond what is possible to reach based on current concrete codes and practice, have been shown to be a very good investment. However, increased and more controlled durability is not only a technical and economic issue, but also an increasingly more important environmental and sustainability issue. Although the present book is mostly concerned with increased and more controlled durability from a technical point of view, a brief introduction to life cycle costs and life cycle assessment is also included.

Acknowledgments

Throughout my work over a span of many years to develop increased and more controlled durability of new important concrete infrastructures, I acknowledge a number of my doctoral students from recent years who have been working with various aspects of concrete durability and contributed to parts of the procedures for both the durability design and the concrete quality control, as outlined and discussed in the present book. These people include Tiewei Zhang, Olaf Lahus, Arne Gussiås, Franz Pruckner, Liang Tong, Surafel Ketema Desta, Miguel Ferreira, Őskan Sengul, Guofei Liu, and Vemund Årskog.

I also thank the Norwegian Coast Directorate and the Norwegian Association for Harbor Engineers for very good research cooperation and support, and in particular I would like to thank Tore Lundestad and Roar Johansen for their great interest and encouragement in trying out and applying the new knowledge to new important concrete infrastructures in Norwegian harbors. As a result of this cooperation, recommendations and guidelines for new durable marine concrete infrastructures were developed and adopted by the Norwegian Association for Harbor Engineers in 2004. Lessons learned from practical applications of these recommendations and guidelines were incorporated into subsequent revised editions, the third and last of which, from 2009, was also adopted by the Norwegian Chapter of PIANC, which is the world association for waterborne transport infrastructure. These recommendations and guidelines are basically the same as those described in the present book, and the DURACON software that provides the basis for the durability analyses is also the same. This software can be freely downloaded from the home page of the Norwegian Chapter of PIANC (http://www.pianc.no/duracon.php).

In this second and revised edition of the current book, more results and experience from practical applications of the above procedures for durability design and concrete quality control applied to recent commercial projects, for both Oslo Harbor KF and Nye Tjuvholmen KS in Oslo, are included. The opportunity to publish all these results is greatly appreciated.

Some preliminary results from the more comprehensive NRF Research Program *Underwater Infrastructure and Underwater City of the Future* at Nanyang Technological University in Singapore are also included, which is greatly appreciated. In this program, the above procedures for durability design and concrete quality control have also been adopted as part of the technical basis for future development of Singapore City based on a large number of sea-spaced concrete structures.

Odd E. Gjørv
Trondheim, Norway

About the Author

Odd E. Gjørv, PhD, DrSc, is professor emeritus in the Department of Structural Engineering at the Norwegian University of Science and Technology (NTNU) in Trondheim, Norway. He joined the Faculty of Technology and Engineering at NTNU in 1971, where he introduced extensive teaching programs in concrete technology at both undergraduate and graduate levels. His teaching also included the supervision of a large number of MSc and PhD students majoring in concrete technology. As a visiting professor, Dr. Gjørv has taught at the University of California, Berkeley, and has given many invited lectures in several countries. He has been a member of the Norwegian Academy of Technical Sciences (NTVA) since 1979 and has

participated in a large number of international professional activities and societies. He is currently engaged as an international collaborator on the NRF Research Program *Underwater Infrastructure and Underwater City of the Future* at Nanyang Technological University in Singapore.

Dr. Gjørv has published more than 350 scientific papers, 2 books, and has contributed to many other professional books. He has received several international awards and honors for his research. He has been a Fellow of the American Concrete Institute since 1989. From 1971 to 1995, he was continuously involved in the development and construction of all the offshore concrete platforms for oil and gas explorations in the North Sea. Dr. Gjørv's research includes advanced concrete materials and concrete construction as well as durability and performance of concrete structures in severe environments. He can be contacted through his website, http://folk.ntnu.no/gjorv/.

Chapter 1

Historical review

When Smeaton constructed the famous lighthouse on Eddystone Rock at the outlet of the English Channel during the period 1756–1759 (Smeaton, 1791), this was the first time a specially developed type of cement for a severe marine environment was applied (Lea, 1970). When the structure was demolished due to severe erosion of the underlying rock in 1877, this structure had remained in very good condition for more than 100 years. Since Smeaton reported his experience on the construction of this lighthouse (Figure 1.1), all the published literature on concrete in marine environments has made up a comprehensive and fascinating chapter in the long history of concrete technology. During the last 150 years, a number of professionals, committees, and national authorities have been engaged in this issue. Numerous papers have been presented to international conferences, such as the International Association for Testing Materials in Copenhagen (1909), New York (1912), and Amsterdam (1927); the Permanent International Association of Navigation Congresses (PIANC) in London (1923), Cairo (1926), Venice (1931), and Lisbon (1949); the International Union of Testing and Research Laboratories for Materials and Structures (RILEM) in Prague in 1961 and 1969; the RILEM-PIANC in Palermo in 1965; and the Fédération Internationale de la Précontrainte (FIP) in Tibilisi in 1972. Already in 1923, Atwood and Johnson (1924) had assembled a list of approximately 3000 references, and still, durability of concrete structures in marine environments continues to be the subject for research, discussion, and international conferences (Malhotra, 1980, 1988, 1996; Mehta, 1989, 1996; Sakai et al., 1995; Gjørv et al., 1998; Banthia et al., 2001; Oh et al., 2004; Toutlemonde et al., 2007; Castro-Borges et al., 2010; Li et al., 2013).

In all this literature, the various deteriorating processes that may affect the durability and performance of concrete structures in severe environments have been extensively reported and discussed. Although a number of deteriorating processes such as alkali–aggregate reactions, freezing and thawing, as well as chemical attack still represent a severe challenge and potential threat to many concrete structures, it is not the disintegration of the concrete itself, but rather chloride-induced corrosion of embedded steel,

1

A

NARRATIVE of the BUILDING

AND

A DESCRIPTION of the CONSTRUCTION

OF THE

EDYSTONE LIGHTHOUSE

WITH STONE:

TO WHICH IS SUBJOINED,

Aᴺ APPENDIX, giving some Account of the LIGHTHOUSE on the SPURN POINT,

BUILT UPON A SAND.

Bʏ JOHN SMEATON, *CIVIL ENGINEER*, F.R.S.

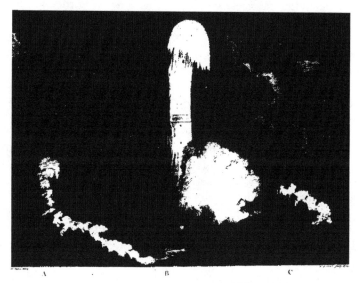

A   B   C

The MORNING after A STORM at S.W.

LONDON:

PRINTED FOR THE AUTHOR, BY H. HUGHS:

SOLD BY G. NICOL,

BOOKSELLER TO HIS MAJESTY, PALL-MALL. 1791.

Figure 1.1 Front page of the report on the construction of the Eddystone Lighthouse, written by John Smeaton in 1791. (Courtesy of the British Museum.)

that appears to be the most severe and greatest threat to the durability and performance of many important concrete structures. Already in 1917, the problem with corrosion of embedded steel was pointed out by Wig and Ferguson (1917) after a comprehensive survey of concrete structures in U.S. waters.

In addition to conventional structures such as bridges and harbor structures, reinforced and prestressed concrete has already, for a long time, been increasingly applied to a large number of very important ocean structures and vessels. Of the total surface area of the globe, ocean water makes up about 70%, and the inhabitable part of the remaining area is even smaller and is becoming increasingly more populated. Since the need for more space, raw materials, and transportation is steadily increasing, increasingly more activities are being moved into ocean waters and marine environments.

Already in the early 1970s, the American Concrete Institute (ACI) came up with a technological forecasting on the future use of concrete, where the rapid development on the continental shelves was pointed out (ACI, 1972). In this report not only structures related to oil and gas explorations but also structures that would relieve land congestion were discussed.

At an international FIP Symposium on Concrete Sea Structures organized by Gosstroy in Tbilisi in 1972 (Gosstroy, 1972), a great variety of concrete structures that would play an increasing role for further activities in ocean and marine environments were discussed. Such structures would be of different types and categories, such as

- Nonanchored freely floating structures, e.g., ships, barges, and containers
- Anchored structures floating at water surface level, e.g., bridges, dry docks, operation platforms, moorings, nuclear plants, airports, and cities
- Anchored structures (positive buoyancy) floating below surface level, e.g., tunnels
- Bottom-supported structures (negative buoyancy) resting above seabed level, e.g., tunnels and storage units
- Bottom-supported structures (negative buoyancy) resting at or below seabed level, e.g., bridges, harbor structures, tunnels, storage units, caissons, operation platforms, as well as both tidal and nuclear power plants

The ACI forecasting pointed out the great potential for utilization of concrete as a construction material for marine and ocean applications in general and for offshore oil and gas exploration in particular. In Norway, where most of the offshore concrete construction has taken place so far, long traditions have existed on the utilization of concrete in the marine environment. Already in the early 1900s, the two Norwegian engineers Gundersen and Hoff developed and obtained a patent on the tremie method for underwater placing of concrete during the construction of the Detroit River Tunnel between the United States and Canada (Gjørv, 1968). From 1910, when

Figure 1.2 Open concrete structures are still the most common type of harbor structures built along the Norwegian coastline.

Gundersen came back to Norway and became the director of the new contracting company AS Høyer-Ellefsen, his newly patented method for underwater placement of concrete became the basis for the construction of a new generation of piers and harbor structures all along the rocky shore of the Norwegian coastline (Gjørv, 1968, 1970). These structures typically consist of an open reinforced concrete deck on top of slender, reinforced concrete pillars cast under water. Although the underwater cast concrete pillars were gradually replaced by driven steel tubes filled with concrete, this open type of concrete structure is still the most common type of harbor structure being constructed along the Norwegian coastline (Figure 1.2).

Due to its very long and broken coastline with many fjords and numerous inhabited islands, Norway has a long tradition on the use of concrete as a construction material in marine environments (Figure 1.3). For many years, this primarily included concrete harbor structures. Gradually, however, concrete also played an increasing role as a construction material for other applications, such as strait crossings (Klinge, 1986; Krokeborg, 1990, 1994, 2001). In addition to conventional bridges (Figure 1.4), new concepts for strait crossings such as floating bridges (Figure 1.5 and 1.6) emerged (Meaas et al., 1994; Hasselø, 2001). Even submerged concrete tunnels have been the subject for detailed studies and planning; one of several types of design is shown in Figure 1.7 (Remseth, 1997; Remseth et al., 1999).

The rapid development that later took place on the utilization of concrete for offshore installations in the North Sea is well known (Figures 1.8 and 1.9). Thus, since 1973, altogether 34 major concrete structures containing more than 2.6 million m³ of high-performance concrete were installed (Figure 1.10), most of which were produced in Norway. Also in other parts of the world, a number of offshore concrete structures have been produced in recent years, and so far, a total of 50 various types of offshore concrete structures have been installed (Moksnes, 2007).

Figure 1.3 Along the Norwegian coastline, with its many deep fjords and numerous small and big islands, there are a large number of both onshore and offshore concrete structures. (Courtesy of NOTEBY AS.)

Figure 1.4 The Tromsø Bridge (1960) is a cantilever bridge with a total length of 1016 m. (Courtesy of Johan Brun.)

Figure 1.5 The Bergsøysund Bridge (1992) is a floating bridge with a total length of 914 m. (Courtesy of Johan Brun.)

For the first offshore concrete platforms in the early 1970s, it was not so easy to produce concrete with the combined requirements of a very low water/cement ratio, high compressive strength, and high amount of entrained air for ensuring proper frost resistance. Extensive research programs were carried out, however, and the quality of concrete and the specified design strength successively increased from project to project

Figure 1.6 The Nordhordlands Bridge (1994) is a combined floating and cable-stayed bridge with a total length of 1610 m. (Courtesy of Johan Brun.)

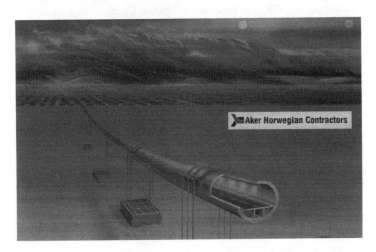

Figure 1.7 One of several designs studied for possible strait crossings by use of submerged tunnels. (Courtesy of the Norwegian Public Roads Administration.)

(Gjørv, 2008). Thus, from the Ekofisk Tank, which was installed in 1973, to the Troll A Platform installed in 1995, the design strength successively increased from 45 to 80 MPa. Also, the water depths for the various installations successively increased. Thus in 1995, the Troll A Platform was installed at a water depth of more than 300 m. From the tip of the skirts to the top of the shafts of this gravity base structure, the total height is 472 m,

Figure 1.8 The first offshore concrete platform, the Ekofisk Tank, on its way out from Stavanger in 1973. (Courtesy of Norwegian Contractors.)

which is taller than the Empire State Building in New York; the artistic view in Figure 1.11 demonstrates the size of the structure. After production in one of the deep Norwegian fjords (Figure 1.12), the Troll A Platform, containing 245,000 m³ of high-strength concrete, 100,000 t of reinforcing steel, and 11,000 t of prestressing steel, was moved out to its final offshore destination, and this operation was the biggest movement of a man-made structure ever made (Figure 1.13). In 1995, the Heidrun Platform was also installed in deep water at a depth of 350 m, but this structure was a tension leg floating platform consisting of lightweight concrete with a design strength of 65 MPa. For the design, detailing, and construction of all these offshore concrete structures, high safety, durability, and serviceability were the subjects of greatest attention and importance.

From the early 1970s, a rapid development on high-strength concrete for both new offshore structures and a number of other types of structures took place (Gjørv, 2008). Since high strength and low porosity also will enhance the overall performance of the material, the term *high-performance concrete* was successively introduced and specified for concrete durability rather than for concrete strength. As experience with this type of concrete was gained, however, it showed that the specification of high-performance concrete was not necessarily good enough for ensuring high durability and service life of concrete structures in severe environments. In recent years, therefore, new and rapid development of more advanced procedures for durability design

Figure 1.9 The Gullfaks C Platform (1989) during construction in Stavanger. (Courtesy of Norwegian Contractors.)

has taken place. As a result, new concrete structures with improved and more controlled durability and service life can now be produced. Thus, for the most exposed parts of the Rion–Antirion Bridge, which was constructed in the southern part of Greece in 2001 (Figure 1.14), concrete with an extremely high resistance to chloride ingress was applied, giving a very high safety against steel corrosion. This type of concrete, which was based on a blast furnace slag cement with a high slag content, showed a 28-day chloride diffusivity of $0.8–1.2 \times 10^{-12}$ m^2/s according to the rapid chloride migration (RCM) method (NORDTEST, 1999), with achieved values for

Figure 1.10 Development of offshore concrete structures in the North Sea. (Courtesy of Aker Solutions.)

the chloride diffusivity on the construction site after one year of $4.0–5.5 \times 10^{-13}$ m^2/s (Kinopraxia Gefyra, 2001).

Before current experience with durability design of new concrete structures is outlined and discussed, it may be useful to take a brief look at and review of current experience with the field performance of existing concrete structures in severe environments.

Figure 1.11 Artistic view of the Troll A Platform (1995) demonstrating that the Oslo City Hall becomes quite small in comparison. (Courtesy of Per Helge Pedersen.)

Figure 1.12 After production in one of the deep Norwegian fjords, the Troll A Platform was ready for towing out to the North Sea in 1995. (Courtesy of Jan Moksnes.)

Figure 1.13 The Troll A Platform (1995) on the way out to its final destination in the North Sea. (Courtesy of Aker Solutions.)

Figure 1.14 The Rion–Antirion Bridge (2001), for which a concrete with an extremely high resistance against chloride ingress was applied. (Courtesy of Gefyra S.A.)

REFERENCES

ACI. (1972). Concrete—Year 2000. *Proceedings ACI Journal*, 68, 581–589.

Atwood, W.G., and Johnson, A.A. (1924). The Disintegration of Cement in Seawater. *Transactions ASCE*, 87, 204–230.

Banthia, N., Sakai, K., and Gjørv, O.E. (eds.). (2001). *Proceedings, Third International Conference on Concrete under Severe Conditions—Environment and Loading*. University of British Columbia, Vancouver.

Castro-Borges, P., Moreno, E.I., Sakai, K., Gjørv, O.E., and Banthia, N. (eds.). (2010). *Proceedings, Sixth International Conference on Concrete under Severe Conditions—Environment and Loading*, vols. 1 and 2. CRC Press, London.

Gjørv, O.E. (1968). *Durability of Reinforced Concrete Wharves in Norwegian Harbours*. Ingeniørforlaget, Oslo.

Gjørv, O.E. (1970). Thin Underwater Concrete Structures. *Journal of the Construction Division, ASCE*, 96, 9–17.

Gjørv, O.E. (2008). High Strength Concrete, in *Developments in the Formulation and Reinforcement of Concrete*, ed. S. Mindess. Woodhead Publishing, pp. 79–97.

Gjørv, O.E., Sakai, K., and Banthia, N. (eds.). (1998). *Proceedings, Second International Conference on Concrete under Severe Conditions—Environment and Loading*. E & FN Spon, London.

Gosstroy. (1972). *Proceedings, FIP Symposium on Concrete Sea Structures*. Moscow.

Hasselø, J.A. (2001). Experiences with Floating Bridges. *Proceedings, Fourth International Symposium on Strait Crossings*, ed. J. Krokeborg. A.A. Balkema Publ., Rotterdam, pp. 333–337.

Kinopraxia Gefyra. (2001). Personal communication.

Klinge, R. (ed.). (1986). *Proceedings, First Symposium on Strait Crossings*. Tapir, Trondheim.

Krokeborg, J. (ed.). (1990). *Proceedings, Second Symposium on Strait Crossings.* A.A. Balkema, Rotterdam.

Krokeborg, J. (ed.). (1994). *Proceedings, Third Symposium on Strait Crossings.* A.A. Balkema, Rotterdam.

Krokeborg, J. (ed.). (2001). *Proceedings, Fourth Symposium on Strait Crossings.* A.A. Balkema, Rotterdam.

Lea, F.M. (1970). *The Chemistry of Cement and Concrete.* Edward Arnold, London.

Li, Z.J., Sun, W., Miao, C.W., Sakai, K., Gjørv, O.E., and Banthia, N. (eds.). (2013). *Proceedings, Seventh International Conference on Concrete under Severe Conditions—Environment and Loading.* RILEM, Bagneux.

Malhotra, V.M. (ed.). (1980). *Proceedings, First International Conference on Performance of Concrete in Marine Environment,* ACI SP-65.

Malhotra, V.M. (ed.). (1988). *Proceedings, Second International Conference on Performance on Concrete in Marine Environment,* ACI SP-109.

Malhotra, V.M. (ed.). (1996). *Proceedings, Third International Conference on Performance on Concrete in Marine Environment,* ACI SP-163.

Meaas, P., Landet, E., and Vindøy, V. (1994). Design of Sahlhus Floating Bridge (Nordhordlands Bridge). In *Proceedings, Third International Symposium on Strait Crossings,* ed. J. Krokeborg. A.A. Balkema Publishing, Rotterdam, pp. 729–734.

Mehta, P.K. (ed.). (1989). *Proceedings, Ben C. Gerwick Symposium on International Experience with Durability of Concrete in Marine Environment.* Department of Civil Engineering, University of California at Berkeley, Berkeley.

Mehta, P.K. (ed.). (1996). *Proceedings, Odd E. Gjørv Symposium on Concrete for Marine Structures.* CANMET/ACI, Ottawa.

Moksnes, J. (2007). Personal communication.

NORDTEST. (1999). *NT Build 492: Concrete, Mortar and Cement Based Repair Materials, Chloride Migration Coefficient from Non-Steady State Migration Experiments.* NORDTEST, Espoo.

Norwegian Public Roads Administration. Private communication.

Oh, B.H., Sakai, K., Gjørv, O.E., and Banthia, N. (eds.). (2004). *Proceedings, Fourth International Conference on Concrete Under Severe Conditions—Environment and Loading.* Seoul National University and Korea Concrete Institute, Seoul.

Remseth, S. (1997). *Proceedings, Analysis and Design of Submerged Floating Tunnels.* Tenth Nordic Seminar on Computional Mechanics, Estonia.

Remseth, S., Leira, B.J., Okstad, K.M., Mathisen, K.M., and Haukås, T. (1999). Dynamic Response and Fluid/Structure Interaction of Submerged Floating Tunnels. *Computers and Structures,* 72, 659–685.

Sakai, K., Banthia, N., and Gjørv, O.E. (eds.). (1995). *Proceedings, First International Conference on Concrete under Severe Conditions—Environment and Loading.* E & FN Spon, London.

Smeaton, J. (1791). *A Narrative of the Building and a Description of the Construction of the Edystone Lighthouse.* H. Hughs, London.

Toutlemonde, F., Sakai, K., Gjørv, O.E., and Banthia, N. (eds.). (2007). *Proceedings, Fifth International Conference on Concrete under Severe Conditions—Environment and Loading.* Laboratoire Central des Ponts et Chauseés, Paris.

Wig, R.J., and Ferguson, L.R. (1917). What Is the Trouble with Concrete in Sea Water? *Engineering News Record,* 79, pp. 532, 641, 689, 737, 794.

Chapter 2

Field experience

2.1 GENERAL

As outlined in Chapter 1, extensive field investigations of a large number of concrete structures in severe environments have been carried out in many countries. For most of these concrete structures, it has primarily been corrosion of embedded steel that has created the most severe problems to the durability and performance. In recent years, the increasing use of de-icing salt has created special problems for many concrete bridges (U.S. Accounting Office, 1979). Already in 1986, it was estimated that the cost of correcting corroding concrete bridges in the United States was US$24 billion, with an annual increase of US$500 million (Transportation Research Board, 1986). Later on, annual costs of repair and replacement of U.S. bridges of up to about US$8.3 billion were estimated by Yunovich et al. (2001), and up to US$9.4 billion for the next 20 years by American Society of Civil Engineers (Darwin, 2007). In 1998, annual costs of US$5 billion for concrete structures in Western Europe were estimated (Knudsen et al., 1998), and similar durability problems and extensive expenses from a large number of other countries have also been reported.

For all concrete structures exposed to marine environments, the environmental conditions may be even more severe (Gjørv, 1975). Thus, along the Norwegian coastline, there are more than 10,000 harbor structures, most of which have been produced by concrete; almost all of them have gotten steel corrosion within a period of about 10 years (Gjørv, 1968, 1994, 1996, 2002, 2006). Also, there are more than 300 large concrete bridges built after 1970 (Figure 2.1), and more than half of them have gotten steel corrosion within a period of about 25 years (Østmoen et al., 1993). In the North Sea, several of the offshore concrete structures have also gotten some extent of steel corrosion, although these concrete structures have generally shown a much better durability.

Internationally, deterioration of major concrete infrastructure has emerged as one of the most severe and demanding challenges facing the construction industry (Horrigmoe, 2000). Although corrosion of embedded

Figure 2.1 The Sortland Bridge (1975) is a 948 m long cantilever bridge in the northern part of Norway. (Courtesy of Johan Brun.)

steel represents the dominating type of deterioration, deterioration due to other processes, such as alkali–aggregate reaction and freezing and thawing, also represents a big problem and challenge to the durability of concrete structures in many countries. In order to describe the field performance of concrete structures in severe environments in more detail, some current experience based on field investigations of concrete structures, mostly in Norwegian waters, is briefly outlined and discussed below.

2.2 HARBOR STRUCTURES

Already in the early 1960s, a broad Nordic research cooperation was established in order to investigate the field performance and service life of concrete structures in Nordic marine environments. In Norway, this work was organized by a technical committee established by the Norwegian Concrete Association, and during the period from 1962 to 1968, extensive field investigations of altogether 219 concrete harbor structures located all along the Norwegian coastline were carried out (Gjørv, 1968; NTNU, 2005). The majority of these concrete harbor structures were of the open type with a reinforced concrete deck on top of slender, tremie-cast concrete pillars (Figures 2.2 and 2.3). The structures had varying ages of up to 50–60 years and included more than 190,000 m² of concrete decks on more than 5,000 tremie-cast concrete pillars, with a total length of approximately 53,000 m.

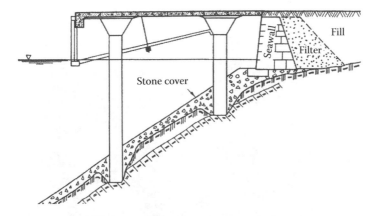

Figure 2.2 A typical section through an open concrete harbor structure with a reinforced concrete deck on top of slender, reinforced concrete pillars. (From Gjørv, O. E., *Durability of Reinforced Concrete Wharves in Norwegian Harbours*, Ingeniørforlaget AS, Oslo, Norway, 1968.)

Figure 2.3 A typical industrial concrete harbor structure. (From Gjørv, O. E., *Durability of Reinforced Concrete Wharves in Norwegian Harbours*, Ingeniørforlaget AS, Oslo, Norway, 1968.)

Of all the concrete structures, more than half of them were also investigated underwater (Figure 2.4).

The overall condition of all the concrete harbor structures investigated was quite good. Even after service periods of up to 50–60 years, the structures still showed a high ability to withstand the combined effects of the most severe marine exposure (Figure 2.5) and heavy structural loads. Thus, one of the industrial harbor structures investigated was observed with storage of raw aluminum bars evenly distributed all over the deck, as shown in

Figure 2.4 The extensive field investigations of Norwegian concrete harbor structures during the 1960s were carried out by H. P. Sundh and O. E. Gjørv.

Figure 2.5 Concrete harbor structures along the Norwegian coastline are exposed to the most severe marine environment. (Courtesy of B. Skarbøvik.)

Figure 2.6; this particular deck load was approximately six times that of the original design load. Apart from severe corrosion in all the deck beams, no special sign of any overloading was observed on this 50-year-old concrete harbor structure.

For the continuously submerged parts of all the structures investigated under water, no particular trend for development of damage, neither due to deterioration of the concrete nor due to corrosion of embedded steel, was

Figure 2.6 A 50-year-old industrial harbor structure (1913) with storage of raw aluminum bars representing an evenly distributed deck load of approximately six times that of the original design load. (From Gjørv, O. E., *Durability of Reinforced Concrete Wharves in Norwegian Harbours*, Ingeniørforlaget AS, Oslo, Norway, 1968.)

observed. Within the tidal zone of the structures, only 35- to 40-year-old structures exhibited some concrete pillars with cross-sectional reductions of more than 20%, mostly due to freezing and thawing (Figure 2.7). Above water, only 35- to 40-year-old structures had deck beams severely weakened due to steel corrosion, while on the whole, both the deck slabs and the rear walls behaved much better.

For 13% of all the structures investigated under water, only some small and confined areas of deteriorated concrete were observed. In these local areas where problems with the tremie placing of the concrete had occurred and given a very porous and permeable concrete, a rapid chemical deterioration of the concrete had taken place. In these local areas, the strength was completely broken down after a relatively short period of time, and the deteriorated concrete was typically characterized by up to 70% loss of lime due to heavy leaching and increased contents of magnesium oxide by up to 10 times that of the original content (Gjørv, 1970). It should be noted that for all these concrete structures, only pure portland cements with medium contents of C_3A had typically been applied, which are known to be quite vulnerable in such aggressive environments.

Already in 1938, extensive long-term field tests on concrete in seawater were started at a field station in Trondheim Harbor. These field tests included more than 2,500 concrete specimens based on 18 different types of commercial cement. After a period of 25–30 years, the different types of cement showed a great variation in the resistance against chemical action

Figure 2.7 Deterioration in the tidal zone mostly due to freezing and thawing. (From Gjørv, O. E., *Durability of Reinforced Concrete Wharves in Norwegian Harbours*, Ingeniørforlaget AS, Oslo, Norway, 1968.)

(a)   (b)

Figure 2.8 Tremie-cast concrete pillars in excellent condition after (a) 34 years of freezing and thawing in Narvik Harbor (1929) and (b) 43 years in Glomfjord Harbor (1920). (From Gjørv, O. E., *Durability of Reinforced Concrete Wharves in Norwegian Harbours*, Ingeniørforlaget AS, Oslo, Norway, 1968.)

of the seawater; the less durable the type of cement, the more important was the porosity and permeability of the concrete (Gjørv, 1971). The best resistance to the chemical action of seawater was observed for the blast furnace slag cements with the highest content of slag, while the least resistance was observed for the pure portland cements, especially for those with the highest contents of C_3A. However, even the low C_3A-containing cements were distinctly affected, but pozzolanic additions significantly improved the resistance to the chemical action of seawater.

Although all the harbor structures along the Norwegian coastline had typically been produced with pure portland cements with medium contents of C_3A, the applied tremie concrete was typically very dense. In order to maintain a proper cohesiveness of the fresh concrete during tremie placing, the concrete was typically produced with a very high cement content of at least 400 kg/m³. Where no dilution of the tremie concrete had taken place during concrete placing, a very good density of the concrete had been obtained, giving very good durability even after exposure periods of up to 50–60 years.

Even after many years of freezing and thawing, the overall condition of the tremie-cast concrete pillars in the tidal zone was also very good. Even in the northern part of Norway, with high amounts of frost and tidal variations of up to 2–3 m, very good conditions, even after 30–40 years of exposure, were observed (Figure 2.8a and b). For those pillars with observed deterioration, the damage was mostly very local, demonstrating a high scatter and variability of achieved concrete quality from one pillar to another (Figure 2.9). For most of the concrete pillars, however, the concrete had been protected by wooden formwork, which typically had been left in place after concreting, but gradually, this formwork had disappeared. Due to the very high density of the tremie concrete, a very good frost resistance had been obtained. It should be noted, however, that most of the concrete structures investigated had been produced at an early period, before any air entraining admixtures were available.

Above water, more than 80% of all the concrete structures had a varying extent of damage due to steel corrosion, and for those structures without any damage, repairs due to steel corrosion had recently been carried out. The first visible sign of steel corrosion in the form of rust staining and cracks typically appeared after a service period of 5–10 years, and it was primarily those parts of the structures that had been the most exposed to intermittent wetting and drying that had been the most vulnerable to corrosion. Typically, this included the lower parts of the deck beams (Figure 2.10) and the rear parts of the concrete decks adjacent to the seawall, where most of the splashing of seawater had taken place (Figure 2.11). When the concrete cover in the lower parts of the deck beams had cracked or spalled off at a very early stage, it was typically observed that the longitudinal bars appeared to be more uniformly corroded, while the shear reinforcement

Figure 2.9 Uneven deterioration in the tidal zone due to high scatter and variability of achieved concrete quality from one pillar to another. (From Gjørv, O. E., *Durability of Reinforced Concrete Wharves in Norwegian Harbours,* Ingeniørforlaget AS, Oslo, Norway, 1968.)

Figure 2.10 The lower parts of the deck beams were typically more vulnerable to steel corrosion than the deck slabs in between. (From Gjørv, O. E., *Durability of Reinforced Concrete Wharves in Norwegian Harbours,* Ingeniørforlaget AS, Oslo, Norway, 1968.)

Figure 2.11 The rear parts of the concrete deck adjacent to the seawall were typi-
cally more vulnerable to steel corrosion than the rest of the deck. (From
Gjørv, O. E., *Durability of Reinforced Concrete Wharves in Norwegian Harbours*,
Ingeniørforlaget AS, Oslo, Norway, 1968.)

and the beam stirrups typically showed a more severe pitting type of corrosion (Figure 2.12).

For the oldest structures, the design strength in the super structure typically varied from 25 to 30 MPa, but gradually, the design strength had
increased up to 35 MPa. The specified minimum concrete covers in the
deck slabs, deck beams, and tremie-cast concrete pillars were typically 25,
40, and 70 mm, respectively. For some of the concrete structures, a cover
thickness of 100 mm in the concrete pillars had also been specified.

Before 1930, experience had typically shown that the deck slabs of the
open concrete decks had performed much better than the deck beams. This
was assumed to be due to an easier and better placing and compaction of
the fresh concrete in the deck slabs compared to that of the deep and narrow
deck beams and girders. The practical consequence of this was drawn in

Figure 2.12 When the concrete cover in the lower parts of the deck beams had cracked or spalled off at a very early stage, the steel bars were more uniformly corroded, while the vertical beam stirrups typically showed a more severe pitting and were often rusted through. (From Gjørv, O. E., *Durability of Reinforced Concrete Wharves in Norwegian Harbours*, Ingeniørforlaget AS, Oslo, Norway, 1968.)

1932, when the first flat type of concrete deck was introduced in Norwegian harbor construction. From then on, a number of structures with a flat type of concrete deck were produced, and these structures showed a much better performance than those with the beam and slab type of deck (Figure 2.13). Since such a structural design was often more expensive, however, the slab and beam type of deck was gradually introduced again. It was assumed that if only the beams were made shallower and wider, it would be equally easy to place and compact the fresh concrete in such structural elements. After some time, however, even the shallower and wider deck beams showed early corrosion, while the flat type of concrete deck without any beams still performed much better.

Figure 2.13 Harbor structures with a concrete deck of the flat type typically showed a much better performance than structures with a beam and slab type of deck. (From Gjørv, O. E., *Durability of Reinforced Concrete Wharves in Norwegian Harbours*, Ingeniørforlaget AS, Oslo, Norway, 1968.)

What was not known in these early days was that concrete structures exposed to a chloride-containing environment after some time would develop a complex system of galvanic cell activities along the embedded steel. In such a system, the more exposed parts of the structure, such as deck beams, would always absorb and accumulate more chlorides, and hence develop anodic areas, while the less exposed parts, such as the slab sections in between, would act as catchments areas for oxygen, and hence form cathodic areas. As a consequence, the more exposed parts of the deck, such as the beams and girders, would always be more vulnerable to steel corrosion than the rest of the concrete deck. This was the reason also why the rear parts of the concrete deck close to the seawall, with more splashing of seawater, would be more vulnerable to steel corrosion than the rest of the concrete deck (Figure 2.11).

The extensive repairs carried out due to steel corrosion in the deck beams also typically showed a very short service life, mostly less than 10 years. Typically, the corroded steel in the locally spalled areas had first been cleaned and then locally patched with new concrete, as shown in Figure 2.14. Also, what was not known in these early days was that such local patch repairs would always set up local changes in the electrolytic conditions, and thus local differences in the electrochemical potentials along the embedded steel. As a result, accelerated corrosion adjacent to the local patched areas would develop, the result of which is typically demonstrated in Figure 2.15. This detrimental effect of patch repairs was first observed and systematically

Figure 2.14 Typical patch repair of a corroded deck beam. (From Gjørv, O. E., *Durability of Reinforced Concrete Wharves in Norwegian Harbours*, Ingeniørforlaget AS, Oslo, Norway, 1968.)

Figure 2.15 Typical accelerated corrosion adjacent to the local patch repair of a deck beam. (From Gjørv, O. E., *Durability of Reinforced Concrete Wharves in Norwegian Harbours*, Ingeniørforlaget AS, Oslo, Norway, 1968.)

investigated and reported on the San Mateo–Hayward Bridge in the early 1950s (Gewertz et al., 1958). Later on, this effect of patch repairs was confirmed by numerous field investigations in many countries.

In spite of the extensive steel corrosion that had been going on in almost all the concrete harbor structures for a long period of time, the effect of this corrosion on the load-bearing and structural capacity of the structures appeared to be rather moderate and slow (Figure 2.16). For each concrete structure investigated, both the effect on structural capacity and extent of damage within the various types of structural elements were rated according

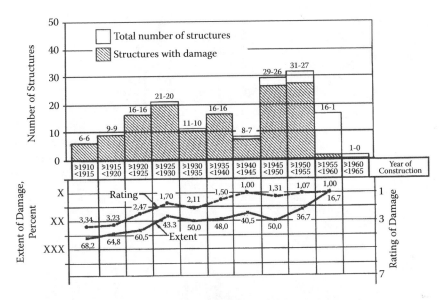

Figure 2.16 Trends for development of deterioration due to steel corrosion in deck beams. (From Gjørv, O. E., *Durability of Reinforced Concrete Wharves in Norwegian Harbours*, Ingeniørforlaget AS, Oslo, Norway, 1968.)

to a given rating system. Thus, the effect on structural capacity was rated according to a scale from 1 to 7, where 1 was distinct observed damage and 3 was major damage but still the elements would fulfill intended function, while 7 was such a severe impairment that the intended function of the elements would no longer be fulfilled. The extent of damage was rated into three groups, where X included up to 1/3 and XXX included more than 2/3 of all the same type of structural elements within each structure with observed damage. As can be seen from Figure 2.16, only 35- to 40-year-old structures (1925–1930) had deck beams with structural ratings of more than 2 and extents of damage of more than 50%.

In 1982–1983, a more detailed field investigation of one of the concrete harbor structures located in Oslo Harbor was carried out (Gjørv and Kashino, 1986). This was a 61-year-old concrete jetty that was going to be demolished in order to create space for new construction work. During demolition, therefore, a unique opportunity occurred for investigating the overall condition of all embedded steel both in the concrete deck and in some of the tremie-cast concrete pillars.

An overall plan of the above jetty is shown in Figure 2.17, which had an open concrete deck of 12,500 m² supported by approximately 300 tremie-cast concrete pillars with cross sections of 90 × 90 cm and heights of up to 17 m. Of all these concrete pillars, four of them were pulled up on shore for a more detailed investigation (Figure 2.18). In 1919–1922, when

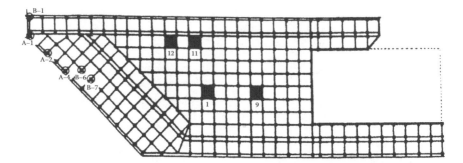

Figure 2.17 An overall plan of the concrete jetty in Oslo Harbor (1922), which was investigated in detail during demolition of the structure after 61 years of service. (From Gjørv, O. E., and Kashino, N., *Materials Performance*, 25(2), 18–26, 1986.)

Figure 2.18 Four of the tremie-cast concrete pillars from the concrete jetty in Oslo Harbor (1922) were pulled up on shore for a very detailed condition assessment of the embedded steel.

the jetty was constructed, the design strength for the superstructure was 25–30 MPa, but at the time of demolition, the in situ compressive strength typically varied from 40 to 45 MPa. Although the original concrete composition was not known, all concrete structures at that early period of construction had typically been produced with a concrete based on very coarse-grained portland cements typically giving a very high strength increase over time. For this particular structure, the specified minimum concrete covers for deck slabs, deck beams, and tremie-cast concrete pillars were 30, 50, and 100 mm, respectively.

Although extensive steel corrosion had been going on in all the deck beams throughout most of the service life of this particular jetty, the overall condition of the structure was quite good even after a service period of more than 60 years. In spite of the extensive corrosion damage and the deep chloride ingress beyond embedded steel throughout the whole concrete deck, the demolition of the deck revealed that most of the rebars were still in quite good condition, practically without any visible corrosion damage. Roughly, this made up more than 75% of the total rebar system. For the rest of the reinforcement, which was mostly located in the lower part of the deck beams, the observed corrosion damage was very unevenly distributed and partly quite severe. However, in the lower parts of the deck beams, the cross section of the bars was seldom reduced by more than 30%, while for the rest of the rebar system, most of the steel bars had reduced cross sections of less than 10%. The best condition of the rebar system was observed in the deck slabs. These observations demonstrate how efficiently the corroding steel in the lower part of the deck beams had functioned as sacrificial anodes, and thus cathodically protected the rest of the rebar system in the deck. This protective effect of the most corroded parts of the rebar system may also explain the relatively slow reduction of structural capacity of all the structures previously investigated (Figure 2.16).

The four tremie-cast concrete pillars that were pulled up on shore for further investigations also showed a very good overall condition. A number of removed concrete cores revealed a compressive strength varying from 40 to 45 MPa. Upon removal of the concrete cover, the embedded steel in the continuously submerged parts of the pillars showed a very good overall condition, mainly due to the low oxygen availability. Above the low-water level, 1 mm deep pittings on the individual bars were typically observed, while below the water level, the pittings were mostly less than 0.2 mm and only occasionally as much as 0.5 mm.

For the concrete jetty as a whole, it was not possible to find any relationship between the half-cell surface potential mapping before demolition and the condition of the embedded steel observed after demolition. The depth of carbonation, which was generally very small, varied according to the prevailing moisture conditions of the concrete. Thus, in the upper parts of the deck with a drier concrete, a carbonation depth of 2–8 mm was typically observed, while for the concrete within the tidal zone and farther below, the carbonation depth was generally very small, typically varying from 1 to 2 mm, and only occasionally as much as 7 mm.

For most of the deck beams, the concrete cover was more or less spalled off, so it was not easy to obtain representative data on the chloride ingress. For the deck slabs, however, the chloride content at the level of the reinforcing bars typically varied from 0.05 to 0.10% by weight of concrete. For the upper part of the pillars above the tidal zone, the chloride content typically varied from 0.15 to 0.25%, while in the tidal zone, the chloride

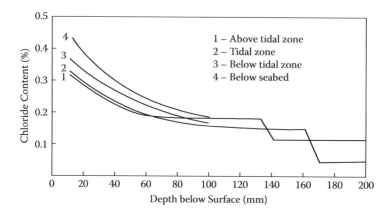

Figure 2.19 Ingress of chlorides in the tremie-cast concrete pillars after 61 years of exposure. (From Gjørv, O. E., and Kashino, N., *Materials Performance*, 25(2), 18–26, 1986.)

content mostly varied from 0.20 to 0.25% (Figure 2.19). For the continuously submerged part of the concrete pillars, an even higher chloride content of 0.30–0.35% was typically observed. As clearly demonstrated in Figure 2.19, the chloride front had reached far beyond the specified cover depth of 100 mm.

In more recent years, concrete qualities, durability specifications, and execution of concrete work for concrete harbor structures have generally improved. However, more recent field investigations of relatively new concrete harbor structures along the Norwegian coastline have revealed that a rapid and uncontrolled chloride ingress still represents a big problem; steel corrosion may still be observed after service periods of less than 10 years. Thus, detailed field investigations of 20 Norwegian concrete harbor structures constructed during the period 1964–1991 showed that 70% had a varying extent of steel corrosion (Lahus et al., 1998; Lahus, 1999). After five years of exposure, average chloride contents at depths of 25 and 50 mm of 0.8 and 0.3% by weight of cement were observed, while after 10 years, the corresponding numbers were 1.2 and 0.5%, respectively. After 15 years of exposure, the average chloride content at a depth of 50 mm was 0.9% by weight of cement.

In Trondheim Harbor, a detailed condition assessment of a cruise terminal (Turistskipskaia) from 1993 showed that steel corrosion had taken place in the concrete deck already after eight years of service (Figure 2.20). Concrete files from the construction period revealed that all requirements according to the then-current concrete codes regarding specified durability had been satisfied (Standard Norway, 1986); concrete with a water/binder ratio of 0.45 and a 28-day compressive strength of 45 MPa had

Figure 2.20 The cruise terminal Turistskipskaia (1993) in Trondheim Harbor with steel corrosion in most of the deck beams after eight years of service. (From Gjørv, O. E., Durability and Service Life of Concrete Structures, in *Proceedings, The First FIB Congress*, 8, 6, Japan Prestressed Concrete Engineering Association, Tokyo, Japan, 2002, pp. 1–16.)

been applied. Also, the concrete had been produced with 380 kg/m^3 high-performance portland cement in combination with 19 kg/m^3 silica fume (5%). Extensive measurements of the achieved concrete cover in the deck beams showed an average value of approximately 50 mm, which was also in accordance with the then-current Norwegian Concrete Code NS 3473 (Standard Norway, 1989).

In spite of meeting all then-current durability requirements, a detailed mapping of electrochemical surface potentials and extensive measurements of the chloride ingress revealed a varying extent of steel corrosion in most of the deck beams; however, no visual damage due to steel corrosion was observed. After approximately eight years of exposure, the chloride front had reached a depth varying from 40 to 50 mm, as typically shown in Figure 2.21. As part of the condition assessment, a number of cores were removed from the deck slabs for testing the chloride diffusivity according to the rapid chloride migration (RCM) method (NORDTEST, 1999). An average value of chloride diffusivity of 10.7 × 10^{-12} m^2/s after eight years in a moist environment only indicated a moderate resistance of the concrete to chloride ingress.

Also at Tjeldbergodden near Trondheim, a deep chloride ingress and ongoing steel corrosion were observed after eight years of exposure in two industrial concrete harbor structures constructed in 1995 and 1996, respectively. Also here, no visual sign of damage was observed, but the

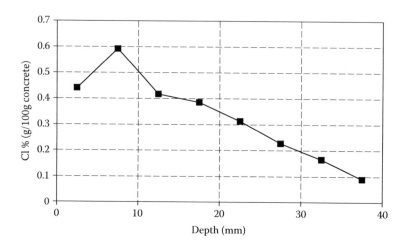

Figure 2.21 Typical chloride ingress in the deck beams of the cruise terminal Turistskipskaia (1993) in Trondheim Harbor after eight years of exposure.

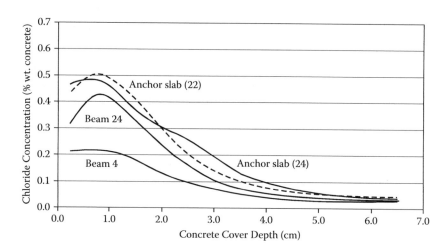

Figure 2.22 Chloride ingress after eight years of exposure in the industrial harbor structure Modulkaia (1995) at Tjeldbergodden. (From Ferreira, M. et al., *Concrete Structures at Tjeldbergodden*, Project Report BML 200303, Department of Structural Engineering, Norwegian University of Science and Technology—NTNU, Trondheim, Norway, 2003.)

chloride front typically varied from 40 to 50 mm (Figure 2.22), and the electrochemical surface potentials revealed corrosion in most of the deck beams. Also for these concrete structures, the specified durability requirements with respect to both concrete quality and concrete cover according to the then-current concrete codes had been fulfilled. During the condition

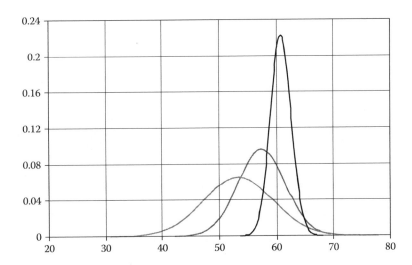

Figure 2.23 Typical variations of concrete cover (mm) in three deck beams from the two industrial harbor structures at Tjeldbergodden. (From Ferreira, M., Probability Based Durability Design of Concrete Structures in Marine Environment, Doctoral Dissertation, Department of Civil Engineering, University of Minho, Guimarães, Portugal, 2004.)

assessment, however, an average chloride diffusivity of 16.6×10^{-12} m²/s (RCM) based on cores from the deck slabs was observed, which also indicates a very low resistance of the concrete to chloride ingress.

Both for the older and newer concrete harbor structures investigated and discussed above, a high scatter and variability of achieved construction quality were typically observed, with respect to both achieved concrete quality and concrete cover. For the two concrete harbor structures at Tjeldbergodden, Figure 2.23 demonstrates the high scatter of achieved concrete cover, typically varying from one deck beam to another.

Also, on another industrial harbor concrete structure produced in 2001 at Ulsteinvik near Ålesund on the Norwegian west coast, a detailed condition assessment typically revealed a high scatter and variability of achieved chloride diffusivity in the concrete deck (Figure 2.24). This testing was based on 12 concrete cores (Ø100 mm) removed from the concrete deck showing a chloride diffusivity varying from 8–9 to 12–13 $\times 10^{-12}$ m²/s (RCM). Although the concrete in question was properly produced by a local ready mix plant in accordance with the current concrete codes, the results in Figure 2.24 clearly demonstrate that the concrete in question only showed a moderate resistance to chloride ingress; the applied water/binder ratio and the minimum cement content were 0.45 and 425 kg/m³ (CEM I), respectively.

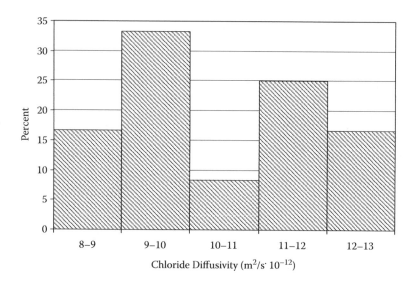

Figure 2.24 Achieved chloride diffusivity in the concrete deck of an industrial concrete harbor structure at Ulsteinvik (2001). (From Guofei, L. et al., *Field Tests with Surface Hydrophobation—Ulsteinvik*, Project Report BML 200503, Department of Structural Engineering, Norwegian University of Science and Technology—NTNU, Trondheim, Norway, 2005.)

For all the concrete harbor structures described and discussed above, it should be noted that the environmental conditions have been quite severe. For certain periods of the year, all the structures had typically been exposed to the most severe combination of splashing seawater and high tides; Figure 2.25 shows the cruise terminal Turistskipskaia in Trondheim Harbor on a stormy day. Occasionally, such structures may be completely submerged in very high tides during stormy periods (Figure 2.26).

For much of the concrete construction work in Norwegian marine environments, the construction takes place all year around. Therefore, the risk of early-age chloride exposure before the concrete has gained sufficient maturity and density is also high. Thus, during concrete construction of the container terminal Nye Filipstadkaia in Oslo Harbor in 2002 (Figure 2.27), the structure was partly exposed to heavy wind and higher tides than normal. As a result, a deep chloride ingress in several of the freshly cast deck beams took place (Figure 2.28). During concrete construction, most types of concrete are very sensitive and vulnerable to chloride exposure; this may represent a special challenge when the concrete is produced with a slow-hydrating binder system during cold weather conditions, which may often be the case in Norwegian marine environments.

In many countries, warm climates with elevated temperatures may also enhance the durability problems due to increased rates of chloride ingress.

Figure 2.25 The cruise terminal Turistskipskaia in Trondheim Harbor during stormy weather. (Courtesy of Trondheim Harbor KS.)

Figure 2.26 Occasionally, concrete harbor structures may be completely submerged in very high tides during stormy periods. (Courtesy of Trondheim Harbor KS.)

Thus, in the Persian Gulf countries, quite extreme durability problems are experienced (Matta, 1993; Alaee, 2000). Similar problems have been reported from a number of other countries with hot climates. This is clearly demonstrated in the Progreso Port on the Yucatán Coast in Mexico, as shown in Figures 2.29–2.32. Due to very shallow waters, two long concrete piers were constructed in order to provide proper harbor facilities, one of which was constructed with traditional reinforcement in the 1960s. Of this

Figure 2.27 During concrete construction of the container terminal Nye Filipstadkaia (2002) in Oslo Harbor, a deep chloride ingress in several of the freshly cast deck beams took place due to heavy wind and high tides. (Courtesy of Oslo Harbor KF.)

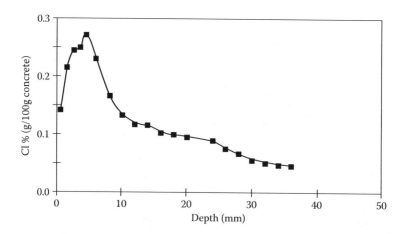

Figure 2.28 Observed chloride ingress in the deck beams during concrete construction of the container terminal Nye Filipstadkaia (2002) in Oslo Harbor. (From Gjørv, O. E., Durability and Service Life of Concrete Structures, in *Proceedings, The First FIB Congress*, 8, 6, Japan Prestressed Concrete Engineering Association, Tokyo, Japan, 2002, pp. 1–16.)

pier, only small parts of the structure still remained in 1998, while the neighboring pier, which was constructed during 1937–1941 with stainless steel reinforcement, was still in very good condition when it was investigated in 1998 (Knudsen and Skovsgaard, 1999). None of the concrete qualities in these two piers was very good. After a service period of approximately

Figure 2.29 Remaining parts of a concrete pier built with black steel reinforcement on the Yucatán Coast in Mexico in the 1960s. (Courtesy of Rambøll Consulting Engineers.)

Figure 2.30 Remaining part of the deck from the concrete pier built on the Yucatán Coast in the 1960s. (Courtesy of Rambøll Consulting Engineers.)

60 years, however, detailed field investigations of the old pier showed that the Ø30 mm stainless steel (AISI 304) was still in very good condition in spite of high chloride contents adjacent to the steel, typically varying from 0.6 to 0.7% by weight of concrete at depths of 80–100 mm below the concrete surface (Rambøll, 1999).

In more recent years, the old Progreso Pier has been extended into deeper waters (Figure 2.33), and it now provides new port facilities for heavy

Figure 2.31 Different durability and long-term performance of the two concrete piers on the Yucatán Coast built with black steel in the 1960s (front) and with stainless steel reinforcement during 1937–1941, respectively. (Courtesy of Rambøll Consulting Engineers.)

Figure 2.32 Progreso Pier on the Yucatán Coast built with stainless steel reinforcement during 1937–1941. (Courtesy of Rambøll Consulting Engineers.)

traffic by a variety of different types of ships (Figure 2.34). As a basis for the design of the old pier in the late 1930s, it was of vital importance for the owner to keep the pier in safe operation with as little future interruption as possible. Therefore, the owner was willing to pay the additional costs for having the structure built with stainless steel reinforcement. Hence, this project clearly demonstrates how the additional costs of stainless steel later on proved to be an extremely good investment for the owner of the structure (Progreso Port Authorities, 2008).

Figure 2.33 Overview of the Progreso Pier with stainless steel reinforcement still in good condition after about 70 years of operation. (Courtesy of Progreso Port Authorities.)

Figure 2.34 The outer part of the Progreso Pier is an important extension of the old concrete pier that was produced with stainless steel reinforcement during 1937–1941. (Courtesy of Progreso Port Authorities.)

2.3 BRIDGES

As already outlined in the introduction to this chapter, extensive corrosion problems have also been experienced for concrete bridges exposed to both de-icing salts and marine environments. Of all the corroding concrete bridges along the Norwegian coastline, one of them was so heavily corroded that it was demolished after a service period of about 25 years (Figure 2.35); this bridge was built in 1970 and belonged to an early generation of Norwegian concrete coastal bridges. During the 25-year service period of this particular bridge, total repair costs comparable to that of the original cost of the bridge had been spent (Hasselø, 1997).

About 10 years later, Gimsøystraumen Bridge (1981) was constructed in the northern part of Norway (Figure 2.36); this is a cantilever bridge that is a more common type of bridge along the Norwegian coastline. However, this bridge also got a deep ingress of chlorides and extensive corrosion after a relatively short period of time. During the repairs of this bridge after 12 years of service, the Norwegian Public Roads Administration selected it as a basis for an extensive research program in order to investigate the effect of various types of patching materials (Blankvoll, 1997).

During this extensive research program that was carried out during the period from 1993 to 1997, it was observed that the deepest chloride ingress had typically taken place in those parts of the bridge that were the least exposed to prevailing winds and salt spray (Figures 2.37 and 2.38). For the most exposed parts of the bridge, rain had intermittently been washing off

Figure 2.35 Ullasundet Bridge (1970) was demolished after 25 years of service due to heavy corrosion of embedded steel. (Courtesy of Jørn A. Hasselø.)

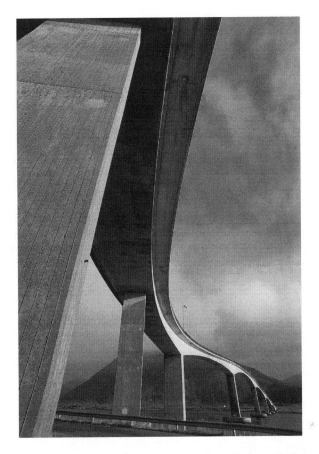

Figure 2.36 Gimsøystraumen Bridge (1981) is a cantilever bridge, which is a very common type of bridge along the Norwegian coastline. (Courtesy of Johan Brun.)

the salt again from the concrete surface, while for the more protected parts and surfaces, the salt had accumulated. The observed chloride ingress also typically varied with height above sea level, as shown in Figure 2.39.

For the superstructure of Gimsøystraumen Bridge, a design strength of 40 MPa had been applied and a minimum concrete cover of 30 mm had been specified. Although such a specified concrete cover was very small for a bridge in a severe marine environment, the achieved concrete cover observed during the extensive repairs was even smaller due to poor workmanship and lack of proper quality control during concrete construction. In Figure 2.40, the results of more than 2028 single measurements of achieved concrete cover are shown.

For the Gimsøystraumen Bridge, the observed moisture content in the outer 40–50 mm of the concrete was typically very high, with relative

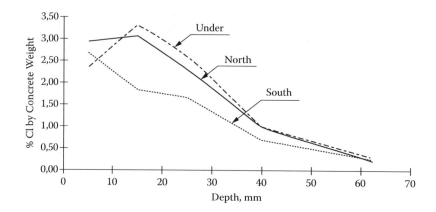

Figure 2.37 Gimsøystraumen Bridge (1981) with deep chloride ingress in the box girder 11.9 m above sea level after 11 years of exposure. (From Fluge, F., Environmental Loads on Coastal Bridges, in *Proceedings, International Conference on Repair of Concrete Structures—From Theory to Practice in a Marine Environment*, ed. A. Blankvoll, Norwegian Public Roads Administration, Oslo, Norway, 1997, pp. 89–98.)

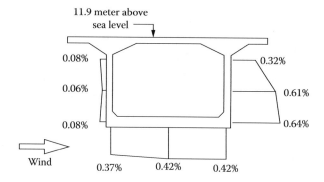

Figure 2.38 Gimsøystraumen Bridge (1981) with the deepest chloride ingress typically observed in those parts of the bridge with the least exposure to prevailing winds and salt spray. (From Fluge, F., Environmental Loads on Coastal Bridges, in *Proceedings, International Conference on Repair of Concrete Structures—From Theory to Practice in a Marine Environment*, ed. A. Blankvoll, Norwegian Public Roads Administration, Oslo, Norway, 1997, pp. 89–98.)

humidities in the range of 70–80% corresponding to a degree of capillary saturation of 80–90% (Sellevold, 1997). Also, for other concrete bridges along the Norwegian coastline, very high moisture contents in the concrete have been observed. Although the moisture contents may vary from one structure to another, typical values for the degree of capillary saturation of 80–90% have been reported (Holen Relling, 1999). For concrete in the

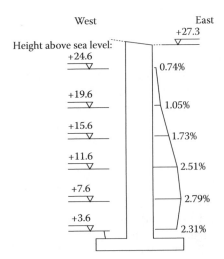

Figure 2.39 Gimsøystraumen Bridge (1981) with typical variation of chloride ingress above sea level. (From Fluge, F., Environmental Loads on Coastal Bridges, in *Proceedings, International Conference on Repair of Concrete Structures—From Theory to Practice in a Marine Environment*, ed. A. Blankvoll, Norwegian Public Roads Administration, Oslo, Norway, 1997, pp. 89–98.)

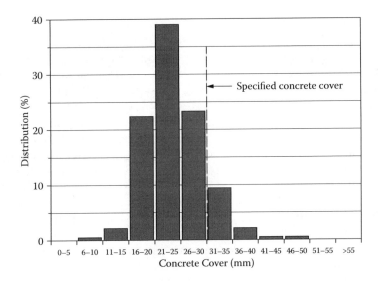

Figure 2.40 Gimsøystraumen Bridge (1981) with observed variation of achieved concrete cover. (From Kompen, R., What Can Be Done to Improve the Quality of New Concrete Structures? in *Proceedings, Second International Conference on Concrete Under Severe Conditions—Environment and Loading*, 3, ed. O. E. Gjørv et al., E & FN Spon, London, England, 1998, pp. 1519–1528.)

tidal and splashing zone, the degree of capillary saturation may be even higher than 90%, while for the more protected parts, the values may be closer to 80%. Thus, for concrete bridges in marine environments typical for the Norwegian coastline, the combination of high chloride contents and high moisture contents appears to give very good conditions for high rates of steel corrosion. In other countries with other climatic conditions, the moisture content in the concrete also may be much lower, or may vary much more over the year, but the temperature may be higher. Although the temperature conditions along the Norwegian coastline may vary, an average annual temperature of 10°C may be typical (Gjørv, 1968).

Already in the early 1950s, extensive field measurements of electrochemical potentials and electrical resistivities along the concrete surface of the San Mateo–Hayward Bridge (1929) in the Bay Area of San Francisco were carried out (Gewertz et al., 1958). For the condition assessment of this particular bridge, equipment such as half-cell copper-copper sulfate electrodes and a four-probe device (Wenner) were applied for the first time reported in the literature (Figure 2.41). During these extensive field investigations, a very close relationship between the ohmic resistance of the concrete and the moisture content of the concrete was observed, and the resistance typically varied from one part of the bridge to another, and also from one period of the year to another. During dry periods when the electrical resistivity of the concrete exceeded a level of about 65,000 ohm cm, a very low and almost negligible rate of corrosion was observed.

The San Mateo–Hayward Bridge had previously also been extensively patch repaired, first by carefully cleaning all damaged areas and then by filling up with shotcrete. Also, for the first time reported in the literature, the detrimental effect of such patch repairs in the form of increased corrosion rates adjacent to the patched areas was systematically investigated and reported (Gewertz et al., 1958).

Also for the Gimsøystraumen Bridge, a continuing heavy steel corrosion was already observed a few years after the extensive patch repairs were finished in 1997. Thus, after 29 years of service, new extensive and costly repairs of the bridge were carried out, but this time, the repairs were based on cathodic protection.

For the younger generation of concrete bridges constructed along the Norwegian coastline, both improved concrete qualities and thicker concrete covers were gradually applied. New concrete bridges with design strengths in the range of 45 to 65 MPa and water/binder ratios of 0.40 or less, in combination with increased concrete covers in the range of 40 to 55 mm, distinctly improved the durability and performance. As already discussed, however, the environmental conditions along the Norwegian coastline can be quite severe (Figure 2.42). Thus, for the Storseisund Bridge, which was built in 1988, deep chloride ingress was observed after approximately 15 years of exposure (Figure 2.43). This bridge is one of several concrete

Figure 2.41 The homemade four-electrode device (Wenner) designed by Richard Stratfull in the early 1950s for electrical resistivity measurements along the concrete surface of the San Mateo–Hayward Bridge (1929). (Courtesy of Richard Stratfull, 1970.)

Figure 2.42 During stormy winter periods, Storseisund Bridge (1988) is heavily exposed to splashing of seawater. (Courtesy of Rolf Jarle Ødegaard.)

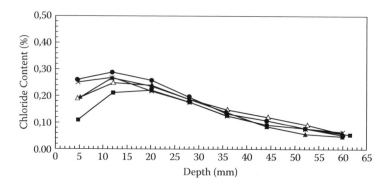

Figure 2.43 Storseisund Bridge (1988) with observed chloride ingress after 15 years of exposure. (From Hasselø, J. A., personal communication, 2007.)

Figure 2.44 During summer, the Atlantic Ocean Road on the Norwegian West Coast is a very pleasant and popular tourist route.

bridges built along the Atlantic Ocean Road on the Norwegian west coast. During summer, this highway is a very pleasant and popular tourist route (Figure 2.44), but during stormy winter periods, these bridges are heavily exposed to splashing of seawater.

In 1991, the 1065 m long cable-stayed Helgelands Bridge (Figure 2.45) was built further north along the Norwegian coastline. During the construction of this bridge, however, the bridge was heavily exposed to severe weather conditions with splashing of seawater already during concrete construction, before the concrete had gained sufficient maturity and density (Figure 2.46a and b). As a result, a deep chloride ingress was already observed shortly after concrete construction (Figure 2.47).

Figure 2.45 The Helgelands Bridge (1991) is a 1065 m long cable-stayed bridge with the largest span of 425 m. (Courtesy of Hallgeir Skog.)

For the Helgelands Bridge, a 45 MPa type of concrete had been applied with an observed in situ strength after two years of 57 MPa; the concrete was produced with 415 kg/m³ cement (CEM I) in combination with 21 kg/m³ silica fume (5%), giving a water/binder ratio of 0.35.

When Aursundet Bridge (1995) was built on the west coast of Norway during 1993–1995 (Figure 2.48), the Norwegian Public Roads Administration selected this bridge as a case study for trying out a new durability design. Thus, on an experimental basis, both a distinctly higher silica fume content in the concrete and a distinctly increased concrete cover in the splash zone were specified. As a result, a concrete based on 400 kg/m³ of cement (CEM I) with 50 kg/m³ of silica fume (12.5%) giving a water/binder ratio of 0.40 was applied. This concrete, which showed a 28-day compressive strength of 55 MPa, was combined with a minimum concrete cover in the splash zone of 80 mm.

After 3 and 10 years of exposure, field investigations revealed an average chloride ingress in the eastern and western parts of the bridge, as shown in Figure 2.49. After 10 years, a testing of the chloride diffusivity (RCM) of the concrete was also carried out, with an average observed value of 6.2×10^{-12} m²/s. Such a level of chloride diffusivity obtained after 10 years in a moist environment only indicates a moderate resistance of the concrete to chloride ingress. However, the observed resistance was much higher than that typically observed for concrete in earlier bridges.

Also, in a number of other countries, extensive field investigations of concrete bridges exposed to marine environments have shown the same type of

(a)

(b)

Figure 2.46 During concrete construction, parts of the Helgelands Bridge (1991) were heavily exposed to severe weather conditions with splashing of seawater. (Courtesy of Hallgeir Skog.)

durability problems due to steel corrosion as those described and discussed for the above bridges along the Norwegian coastline (Malhotra, 1980, 1988, 1996; Mehta, 1989, 1996; Nilsson, 1991; Stoltzner and Sørensen, 1994; Sakai et al., 1995; Beslac et al., 1997; Wood and Crerar, 1997; Gjørv et al., 1998; Banthia et al., 2001; Oh et al., 2004; Toutlemonde et al., 2007; Castro-Borges et al., 2010; Li et al., 2013).

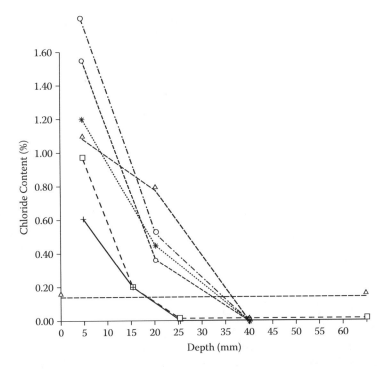

Figure 2.47 The Helgelands Bridge (1991) with observed chloride ingress shortly after concrete construction. (From NPRA, *The Helgelands Bridge—Chloride Penetration*, Internal Report, Norwegian Public Road Administration— NPRA, Oslo, Norway, 1993 [in Norwegian].)

Figure 2.48 The Aursundet Bridge (1995) is a cantilever bridge with a total length of 486 m.

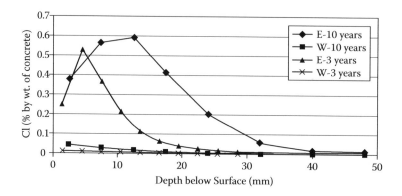

Figure 2.49 The Aursundet Bridge (1995) with observed chloride ingress after 3 and 10 years of splash zone exposure. (From Årskog, V. et al., Chloride Penetration into Silica Fume Concrete after 10 Years of Exposure in Aursundet Bridge, in *Proceedings, Nordic Concrete Research Meeting*, ed. T. Kanstad and E. A. Hansen, Norwegian Concrete Association, Oslo, Norway, 2005, pp. 97–99.)

2.4 OFFSHORE STRUCTURES

Since the early 1970s, altogether 34 major concrete structures for the oil and gas explorations in the North Sea have been installed. In spite of the very harsh and hostile marine environment (Figures 2.50 and 2.51), the overall durability and performance of these concrete structures have shown to be very good (Fjeld and Røland, 1982; Hølaas, 1992; Gjørv, 1994; FIP, 1996; Moksnes and Sandvik, 1996; Helland et al., 2010).

In spite of the overall good durability and performance, however, a certain amount of very costly repairs due to corrosion of embedded steel has been carried out. Thus, after 13 years of service, extensive repairs based on cathodic protection of the Oseberg A Platform (1988) were carried out (Figures 2.52 and 2.53). For this particular structure, the achieved concrete cover was distinctly less than that specified in the upper parts of the shafts (Østmoen, 1998). A nominal concrete cover of 75 mm had been specified, but the achieved concrete cover was highly variable and partly very low. For the upper parts of the shafts, a 60 MPa type of concrete with a water/cement ratio of 0.37 (CEM I) had been applied.

Although no systematic monitoring of the chloride ingress for any of the offshore concrete structures has been carried out, a certain amount of field investigations has revealed that also for these concrete structures, a certain rate of chloride ingress has taken place. Thus, typical chloride ingress in the Heidrun Platform (1995) after two years of exposure is shown in Figure 2.54. This is a floating tension leg platform produced with high-performance lightweight concrete. With a design strength of 60 MPa, this

Figure 2.50 The Frigg TCP 2 Platform (1977) in stormy weather.

concrete was produced with a cement of type CEM I in combination with 5% silica fume at a water/binder ratio of 0.39. Above water, the structure was partly protected by a thin epoxy coating, but as can be seen from Figure 2.54, this particular coating had not been very efficient in keeping the chlorides out.

After eight years of exposure of the Statfjord A Platform (1977), the chloride ingress 7 m above water is shown in Figure 2.55. With a design strength of 50 MPa, this concrete had been produced with a CEM I type of cement at a water/cement ratio of 0.38. From Figure 2.55 it can also be seen that those parts of the structure that had been protected by a solid epoxy coating in the tidal and splash zone did not have any chloride ingress during a service period of eight years. For most of the concrete structures produced before 1980, a solid epoxy coating 2–3 mm thick had typically been applied as an additional protection of the structures above water. Since this protective coating had been continuously applied during slip forming when the young concrete still had an underpressure and suction ability, a very good bonding between the concrete substrate and the coating was achieved. Even after 15 years of exposure, later investigations have revealed that this protective coating is still intact and has very effectively prevented any chlorides from penetrating the concrete (Årstein et al., 1998).

After 17 years of exposure, typical ingress of chlorides in the Ekofisk Tank is shown Figure 2.56. With a design strength of 45 MPa, this concrete had been produced with an ordinary portland cement (CEM I) at a water/cement ratio of 0.45. In the early 1970s, it was not so easy to produce

Figure 2.51 All the concrete platforms in the North Sea are exposed to heavy wave loading and splashing of seawater.

a concrete with strict requirements to both high compressive strength and high air-void contents for frost resistance. However, the Ekofisk Tank is the only platform in the North Sea that was produced with a concrete for the tidal and splashing zone at a water/binder ratio above 0.40.

Although the Brent B Platform (1975) was also one of the earliest platforms produced for the North Sea, the concrete applied for the tidal and splash zone of this structure had a water/cement ratio of 0.40, in combination with a cement content of more than 400 kg/m^3 (CEM I). During concrete construction, the extensive quality control of the air-entrained concrete above water and the non-air-entrained concrete below water showed average 28-day compressive strengths of 48.5 and 56.9 MPa, respectively. Above water, a nominal concrete cover of 75 mm was specified, and this cover was secured by use of mortar blocks of strength and durability comparable to that of the structural concrete. For this particular platform, however, no protective surface coating above water was applied.

Figure 2.52 The Oseberg A Platform (1988). (Courtesy of Trond Østmoen.)

Figure 2.53 After 13 years of service, extensive and very costly repairs based on cathodic protection of the Oseberg A Platform (1988) were carried out. (Courtesy of Trond Østmoen.)

In conjunction with extensive installation work carried out on the Brent B Platform in 1994, a large number of deep Ø100 mm concrete cores were removed from the utility shaft at two different elevations above water and one elevation below water. Therefore, a unique opportunity occurred to investigate the outer part of all these concrete cores for chloride ingress both above and below water after approximately 20 years of exposure (Sengul and Gjørv, 2007). Further below the concrete surface layer of the cores, other properties of the concrete were also investigated.

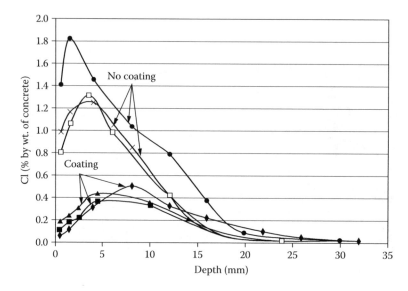

Figure 2.54 Chloride ingress in the Heidrun Platform (1995) after two years of exposure. (From Gjørv, O. E., Durability and Service Life of Concrete Structures, in *Proceedings, The First FIB Congress*, 8, 6, Japan Prestressed Concrete Engineering Association, Tokyo, Japan, 2002, pp. 1–16.)

As can be seen from Figures 2.57–2.60, a deep chloride ingress had taken place, with the deepest ingress in the upper part of the splash zone and the lowest ingress in the constantly submerged part of the shaft at an elevation of –11.5 m. In the upper part of the shaft, at approximately 14 m above water level, a chloride front of approximately 0.07% by weight of concrete at a depth of approximately 60 mm was observed. For a nominal concrete cover of 75 mm specified, this indicates that an early stage of depassivation had probably been reached.

Further below the surface layer of the concrete cores, both the quality and the homogeneity of the concrete were tested by use of the RCM method (NORDTEST, 1999). These tests were carried out in various depths from the concrete surface on altogether 14 Ø100 mm cores from all three levels of the concrete shaft. As can be seen from Figure 2.61, the observed chloride diffusivity varied from approximately 18–20 up to 32–34 × 10^{-12} m^2/s. Although great efforts had been made in order to produce a concrete as homogeneous as possible, the results in Figure 2.61 clearly demonstrate the high scatter and variability of achieved concrete quality. The levels of observed chloride diffusivity after approximately 20 years of curing in

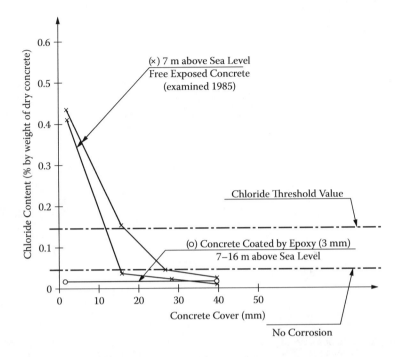

Figure 2.55 Chloride ingress in the Statfjord A Platform (1977) after eight years of exposure. (From Sandvik, M., and Wick, S. O., Chloride Penetration into Concrete Platforms in the North Sea, in *Proceedings, Workshop on Chloride Penetration into Concrete Structures*, ed. L.-O. Nilsson, Division of Building Materials, Chalmers University of Technology, Gothenburg, Sweden, 1993.)

a moist environment also demonstrate a relatively low resistance of the concrete to chloride ingress.

In 1994, a number of concrete cores were also removed from below water of the Brent C Platform (1978). After approximately 17 years of exposure, investigations of these concrete cores from elevations of −9 to −18.5 m also revealed a deep and varying chloride ingress, as shown in Figure 2.62. The high scatter of observed chloride ingress in both the Brent B and Brent C platforms may also reflect the high scatter and variability of achieved concrete quality.

For the production of all the offshore concrete platforms, great efforts were made in order to produce a highest possible homogeneity of the concrete. Thus, from the regular concrete quality control during construction of all the concrete platforms produced during the period 1972–1984, the standard deviation from the 28-day testing of compressive strength typically varied from 2.3 to 3.9 MPa, as shown in Table 2.1.

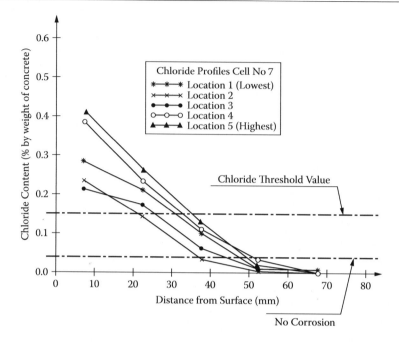

Figure 2.56 Chloride ingress in the Ekofisk Tank (1973) after 17 years of exposure. (From Sandvik, M. et al., *Chloride Permeability of High-Strength Concrete Platforms in the North Sea*, ACI SP-145, ed. V. M. Malhotra, ACI, Detroit, Michigan, 1994, pp. 121–130.)

Table 2.1 Concrete quality control based on 100 mm cubes from platform construction during the period 1972–1984

	28-day compressive strength (MPa)			
Platform (year)	Specified grade	Obtained mean	Standard deviation	Obtained grade[a]
Ekofisk I (1972)	40[b]	45[b]	2.3[b]	41.6[b]
		57	3.5	51.9
Beryl A (1974)	45	55	3.0	50.7
Brent B (1974)	45	53	3.1	48.5
Brent D (1975)	50	54.2	2.5	50.6
Statfjord A (1975)	50	54.6	3.0	50.2
Statfjord B (1979)	55	62.5	3.9	56.9
Statfjord C (1982)	55	67.5	3.8	62.0
Gullfaks A (1984)	55	65.2	3.3	60.3

Source: Moksnes, J., and Jakobsen, B., *High-Strength Concrete Development and Potentials for Platform Design*, OTC Paper 5073, Annual Offshore Technology Conference, Houston, Texas, 1985, pp. 485–495.

[a] Obtained grade = obtained mean − 1.45 × standard deviation.
[b] 150 × 300 mm cylinder.

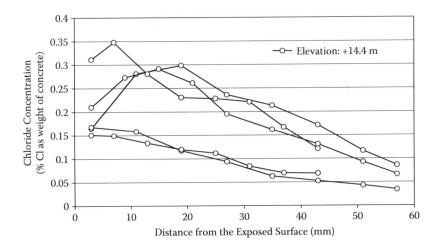

Figure 2.57 The Brent B Platform (1975) with observed chloride ingress at elevation + 14.4 m above water after 20 years of exposure. (From Sengul, Ő., and Gjørv, O. E., Chloride Penetration into a 20 Year Old North Sea Concrete Platform, in *Proceedings, Fifth International Conference on Concrete under Severe Conditions—Environment and Loading,* ed. F. Toutlemonde et al., vol. I, Laboratoire Central des Ponts et Chaussées, Paris, France, 2007, pp. 107–116.)

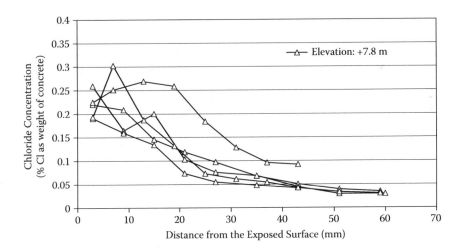

Figure 2.58 The Brent B Platform (1975) with observed chloride ingress at elevation + 7.8 m above water after 20 years of exposure. (From Sengul, Ő., and Gjørv, O. E., Chloride Penetration into a 20 Year Old North Sea Concrete Platform, in *Proceedings, Fifth International Conference on Concrete under Severe Conditions—Environment and Loading,* ed. F. Toutlemonde et al., vol. I, Laboratoire Central des Ponts et Chaussées, Paris, France, 2007, pp. 107–116.)

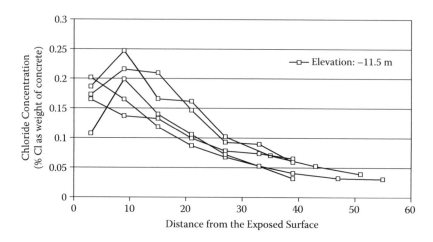

Figure 2.59 The Brent B Platform (1975) with observed chloride ingress at elevation −11.5 m below water after 20 years of exposure. (From Sengul, Ő., and Gjørv, O. E., Chloride Penetration into a 20 Year Old North Sea Concrete Platform, in *Proceedings, Fifth International Conference on Concrete under Severe Conditions—Environment and Loading*, ed. F. Toutlemonde et al., vol. I, Laboratoire Central des Ponts et Chaussées, Paris, France, 2007, pp. 107–116.)

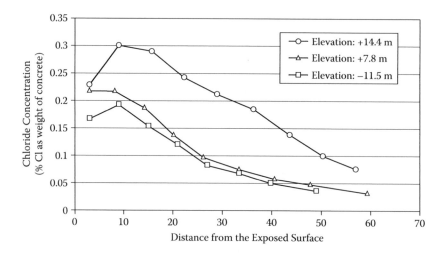

Figure 2.60 The Brent B Platform (1975) with observed chloride ingress after 20 years of exposure. (From Sengul, Ő., and Gjørv, O. E., Chloride Penetration into a 20 Year Old North Sea Concrete Platform, in *Proceedings, Fifth International Conference on Concrete under Severe Conditions—Environment and Loading*, ed. F. Toutlemonde et al., vol. I, Laboratoire Central des Ponts et Chaussées, Paris, France, 2007, pp. 107–116.)

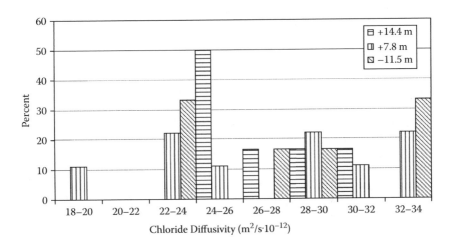

Figure 2.61 Observed chloride diffusivity in the utility shaft of the Brent B Platform (1975). (From Årskog, V., and Gjørv, O. E., Unpublished results, Department of Structural Engineering, Norwegian University of Science and Technology— NTNU, Trondheim, Norway, 2008.)

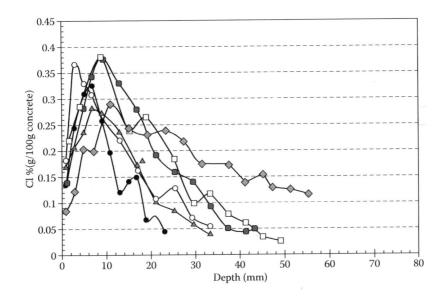

Figure 2.62 The Brent C Platform (1978) with observed chloride ingress below water after 17 years of exposure. (From Gjørv, O. E., Durability and Service Life of Concrete Structures, in *Proceedings, The First FIB Congress*, 8, 6, Japan Prestressed Concrete Engineering Association, Tokyo, Japan, 2002, pp. 1–16.)

In order to further improve the homogeneity of the applied concrete for platform construction, a new type of industrial processing of the fine concrete aggregate was introduced in the early 1980s (Moksnes, 1982). In this processing, all the fine aggregate (0–5 mm) was sent through a water tank and separated by flotation in eight size ranges before it was automatically put together again in order to obtain a very constant and optimal grading curve. As a result, an even more homogeneous concrete was produced for all the other concrete platforms.

From Table 2.1 it can be seen that the concrete produced for the Brent B Platform typically showed a standard deviation of 3.1 MPa. In spite of such a homogeneous concrete, a very high scatter and variability of achieved concrete quality was observed (Figure 2.61). Due to the very densely reinforced structures of up to more than 600 kg steel per m^3 of concrete and a very fluid type of concrete, it was very difficult to avoid some extent of segregation and inhomogeneity during placing and compaction of the concrete; the maximum size of the aggregate was typically 20 mm. Therefore, also newer concrete platforms typically showed a certain scatter and variability of achieved concrete quality. Occasionally it was also very difficult to keep the specified concrete cover.

2.5 OTHER STRUCTURES

In addition to the field performance of the various categories of concrete structures as outlined above, there are also a variety of other types of concrete structures showing problems due to an uncontrolled durability and service life. In the marine environment, there are a number of buildings and other facilities that are also suffering from chloride-induced corrosion. Also, in addition to all the corroding highway bridges exposed to de-icing salt, there are also a large number of parking garages that are suffering from severe corrosion problems due to the contamination of de-icing salt, the consequences of which have occasionally been quite severe (Figure 2.63).

When the cars bring in de-icing salt solution from the outside, and thus contaminate the concrete decks in parking garages, a pattern of corrosion activities in the parking bays is often observed, as typically shown in Figure 2.64. Such a corrosion activity mapping that is the combined result of potential and resistivity measurements along the concrete deck has proved to be a very efficient tool for condition assessment of existing concrete structures in severe environments (Pruckner, 2002; Pruckner and Gjørv, 2002).

Figure 2.63 The consequences of rebar corrosion in parking garages exposed to de-icing salt contamination have occasionally been quite severe. (From Simon, P., Improved Current Distribution due to a Unique Anode Mesh Placement in a Steel Reinforced Concrete Parking Garage Slab CP System, in *NACE Corrosion 2004*, Paper 04345, 2004.)

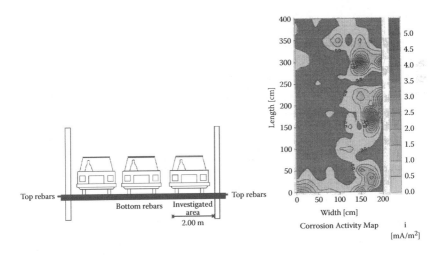

Figure 2.64 Typical pattern of corrosion activities in the parking bays of a parking garage contaminated with de-icing salt. (From Pruckner, F., Diagnosis and Protection of Corroding Steel in Concrete, PhD Thesis 2002:140, Department of Structural Engineering, Norwegian University of Science and Technology—NTNU, Trondheim, Norway, 2002.)

2.6 SUMMARY

2.6.1 General

It should be noted that most of the concrete structures described above have been exposed to climatic conditions with typically low to moderate temperatures. Since most of the mechanisms for degradation of concrete structures are very temperature dependent, however, concrete structures exposed to other climatic conditions with more elevated temperatures may therefore perform quite differently. Also, almost all of the above concrete structures have typically been produced with pure portland cements (CEM I), while concrete structures produced with other types of binder system may also show a different durability and performance. Thus, extensive experience demonstrates that concrete structures produced with blast furnace slag cements typically show a distinctly better performance in chloride-containing environments than those produced with pure portland cements (Bijen, 1998). From all the above concrete structures, however, the field experience can be briefly summarized as shown below.

2.6.2 Deteriorating mechanisms

2.6.2.1 Corrosion of embedded steel

For all the above concrete structures, chloride-induced corrosion of embedded steel was the dominating and most severe problem to the durability and performance; concrete carbonation was not a problem for the observed steel corrosion. For new concrete structures in severe environments, therefore, carbonation-induced corrosion should not be of any special concern to the durability design. For concrete structures in warmer climates, however, much higher rates of chloride ingress must be expected compared to that typically observed for the above structures. For new concrete structures in warmer climates, therefore, a proper control of the chloride ingress may represent an even bigger challenge both to the durability design and to the operation of the concrete structures.

2.6.2.2 Alkali–aggregate reaction

Since most of the above concrete structures were produced in Norway, it should be noted that also in this country alkali–aggregate reaction (AAR) represents a certain problem to the durability of concrete structures. Apart from a very few cases of observed damage, however, deterioration of the concrete due to AAR was not a special problem to the durability of the above structures.

In certain countries, however, AAR may still represent a big problem and challenge to the durability and performance of many important concrete structures. In recent years, however, much international experience has been gained in improved procedures and guidelines for selecting proper qualities of both concrete aggregate and binder systems for control of AAR.

2.6.2.3 Freezing and thawing

Although deterioration of the concrete due to freezing and thawing also was a potential problem for most of the above concrete structures, only a small amount of such damage was observed; the damage was mostly confined to local areas, where the concrete structures had typically gotten a high scatter and variability of achieved concrete quality.

Many of the above concrete harbor structures along the Norwegian coastline were constructed in an early period before any air entraining admixtures were available. Due to the very dense tremie concrete typically applied in the tidal zone of all these structures, however, a very good frost resistance, even after 30 to 40 years of exposure, was obtained.

Even for concrete based on blast furnace slag cements, which is generally more susceptible to frost scaling, general experience shows that good frost resistance is obtained if the concrete is made sufficiently dense, e.g., water/binder ratios ≤ 0.40. Thus, at the Treat Island Field Station on the Canadian east coast, concrete based on slag cements with up to 80% slag performed very well after 25 years of very severe freezing and thawing in the tidal zone; this concrete was produced with a water/binder ratio of 0.40 (Thomas et al., 2011).

2.6.3 Codes and practice

For the above offshore concrete structures, a service life of 30 years was typically specified for a long time. However, as the technology for oil and gas production gradually improved and the service life of the oil and gas fields gradually increased, the specified service life for new offshore concrete structures also gradually increased up to typically 60 years. For all the land-based concrete structures produced along the Norwegian coastline during the same period, however, the specified service life typically was 60 years, gradually increasing up to 100 years. In spite of this difference in required service life, significantly stricter durability requirements were specified for all the offshore concrete structures compared to those for all the land-based marine concrete structures.

In order to meet the strict requirements from the very demanding offshore industry for use of structural concrete for installations in the North

Sea, the International Organization for Prestressed Concrete Structures (FIP) produced some special recommendations for the design and construction of concrete sea structures in 1973 (FIP, 1973). According to the durability requirements in these recommendations, the water/cement ratio should not exceed 0.45 for the most exposed parts of the structures in combination with a minimum cement content of 400 kg/m³, but preferably, the water/cement ratio should not exceed 0.40. For the most exposed parts of the structures, the nominal concrete cover to principal reinforcement and to prestressing tendons should also be 75 and 100 mm, respectively. Already in 1976, these recommendations for concrete sea structures were adopted by both the Norwegian Petroleum Directorate (NPD, 1976) and Det Norske Veritas (DNV, 1976) for offshore concrete structures in the North Sea. To a great extent, the durability requirements in the above FIP recommendations were based on the conclusions and recommendations from the comprehensive field investigations of concrete harbor structures carried out along the Norwegian coastline during the period 1962–1968 (Gjørv, 1968).

In the early 1970s, it was not easy to produce concrete with a water/cement ratio of 0.40 or lower. For the Ekofisk Tank, which was the first concrete platform installed in 1973, a concrete for the tidal and splash zone with a water/binder ratio of 0.45 was produced, while for all the other concrete platforms produced later on, the water/binder ratio typically varied between 0.35 and 0.40 in combination with a cement content of more than 400 kg/m³.

As will be discussed in more detail in Chapter 12, the durability requirements for all the land-based concrete structures produced along the Norwegian coastline were lacking far behind current knowledge and state of the art. Thus, in the early 1970s, the Norwegian Concrete Code did not have any requirement to the water/cement ratio for concrete structures in marine environments, while the requirement to minimum concrete cover was 25 mm (Standard Norway, 1973).

While it only took five years before FIP adopted the above conclusions and recommendations from the comprehensive field investigations of Norwegian concrete harbor structures carried out during the 1960s, it took 18 years before the Norwegian Concrete Code introduced a water/cement ratio requirement of ≤0.45 for the tidal and splashing zone of concrete structures in marine environments (Standard Norway, 1986). Also, it took 35 years before the requirement of a water/cement ratio of ≤0.40 was adopted (Standard Norway, 2003a). After 21 and 35 years, the minimum concrete cover was also increased from 25 to 50 and 60 mm, respectively (Standard Norway, 1989 and 2003b). Still, many of the current European Concrete Codes only require a water/cement ratio of ≤0.45 for the most exposed parts of concrete structures in marine environments (CEN, 2009).

2.6.4 Achieved construction quality

For all the above concrete structures, both the achieved concrete quality and concrete cover typically showed a high scatter and variability, and in the above severe environments, any weaknesses and deficiencies had soon been revealed, whatever durability specifications and materials had been applied. Even for the offshore concrete structures, where very great efforts and strict procedures for both concrete production and concrete quality control were applied, these structures also partly showed a very high scatter and variability of achieved construction quality.

2.6.5 Operation of the structures

For all the above concrete structures, the typical situation for the operation of the structures had been that the maintenance and repairs had typically been reactive, and the need for taking appropriate measures had typically been realized at an advanced stage of degradation. For chloride-induced corrosion of embedded steel, repairs at such a stage are then both technically difficult and disproportionally costly compared to that of carrying out preventive maintenance based on regular condition assessments.

REFERENCES

Alaee, M.J. (ed.). (2000). *Proceedings, Fourth International Conference on Coasts, Ports and Marine Structures—ICOPMAS 2000.* Iranian Ports and Shipping Organization—PSO, Teheran.

Årskog, V., and Gjørv, O.E. (2008). Unpublished Results. Department of Structural Engineering, Norwegian University of Science and Technology—NTNU, Trondheim.

Årskog, V., Gjørv, O.E., Sengul, Ő., and Dahl, R. (2005). Chloride Penetration into Silica Fume Concrete after 10 Years of Exposure in Aursundet Bridge. In *Proceedings, Nordic Concrete Research Meeting*, ed. T. Kanstad and E.A. Hansen. Norwegian Concrete Association, Oslo, pp. 97–99.

Årstein, R., Rindarøy, O.E., Liodden, O., and Jenssen, B.W. (1998). Effect of Coatings on Chloride Penetration into Offshore Concrete Structures. In *Proceedings, Second International Conference on Concrete under Severe Conditions—Environment and Loading*, vol. 2, ed. O.E. Gjørv, K. Sakai, and N. Banthia. E & FN Spon, London, pp. 921–929.

Banthia, N., Sakai, K., and Gjørv, O.E. (eds.). (2001). *Proceedings, Third International Conference on Concrete under Severe Conditions—Environment and Loading.* University of British Columbia, Vancouver.

Beslac, J., Hranilovic, M., Maric, Z., and Sesar, P. (1997). The Krk Bridge: Chloride Corrosion and Protection. In *Proceedings, International Conference on Repair of Concrete Structures—From Theory to Practice in a Marine Environment*, ed. A. Blankvoll. Norwegian Public Roads Administration, Oslo, pp. 501–506.

Bijen, J. (1998). *Blast Furnace Slag for Durable Marine Structures*. VNC/BetonPrisma, Hertogenbosch.

Blankvoll, A. (ed.). (1997). *Proceedings, Repair of Concrete Structures—From Theory to Practice in a Marine Environment*. Norwegian Road Research Laboratory, Oslo.

Castro-Borges, P., Moreno, E.I., Sakai, K., Gjørv, O.E., and Banthia, N. (eds.). (2010). *Proceedings Vol. 1 and 2, Sixth International Conference on Concrete under Severe Conditions—Environment and Loading*. CRC Press, London.

CEN. (2009). *Survey of National Requirements Used in Conjunction with EN 206-1:2000*, Technical Report CEN/TR 15868. CEN, Brussels.

Darwin, D. (2007). President's Memo: It's Time to Invest. *Concrete International*, 29(10), 7.

DNV. (1976). *Rules for the Design, Construction and Inspection of Fixed Offshore Structures*. Det Norske Veritas—DNV, Oslo.

Ferreira, M. (2004). Probability Based Durability Design of Concrete Structures in Marine Environment, Doctoral Dissertation. Department of Civil Engineering, University of Minho, Guimarães.

Ferreira, M., Årskog, V., and Gjørv, O.E. (2003). *Concrete Structures at Tjeldbergodden*, Project Report BML 200303. Department of Structural Engineering, Norwegian University of Science and Technology—NTNU, Trondheim.

FIP. (1973). *Recommendations for the Design and Construction of Concrete Sea Structures*. Féderation Internationale de la Précontrainte—FIP, London.

FIP. (1996). *Durability of Concrete Structures in the North Sea*, State-of-the-Art Report. Féderation Internationale de la Précontrainte—FIP, London.

Fjeld, S., and Røland, B. (1982). Experience from In-Service Inspection and Monitoring of 11 North Sea Structures. In *Offshore Technology Conference*, Paper 4358, Houston.

Fluge, F. (1997). Environmental Loads on Coastal Bridges. In *Proceedings, International Conference on Repair of Concrete Structures—From Theory to Practice in a Marine Environment*, ed. A. Blankvoll. Norwegian Public Roads Administration, Oslo, pp. 89–98.

Gewertz, M.W., Tremper, B., Beaton, J.L., and Stratfull, R.F. (1958). *Causes and Repair of Deterioration to a California Bridge due to Corrosion of Reinforcing Steel in a Marine Environment*, Highway Research Board Bulletin 182, National Research Council Publication 546. National Academy of Sciences, Washington, DC.

Gjørv, O.E. (1968). *Durability of Reinforced Concrete Wharves in Norwegian Harbours*. Ingeniørforlaget AS, Oslo.

Gjørv, O.E. (1970). Thin Underwater Concrete Structures. *Journal of the Construction Division*, ASCE, 96, 9–17.

Gjørv, O.E. (1971). Long-Time Durability of Concrete in Seawater. *Journal of American Concrete Institute*, 68(1), 60–67.

Gjørv, O.E. (1975). Concrete in the Oceans. *Marine Science Communications*, 1(1), 51–74.

Gjørv, O.E. (1994). Steel Corrosion in Concrete Structures Exposed to Norwegian Marine Environment. *Concrete International*, 16(4), 35–39.

Gjørv, O.E. (1996). Performance and Serviceability of Concrete Structures in the Marine Environment. In *Proceedings, Odd E. Gjørv Symposium on Concrete for Marine Structures*, ed. P.K. Mehta. ACI/CANMET, Ottawa, pp. 259–279.

Gjørv, O.E. (2002). Durability and Service Life of Concrete Structures. In *Proceedings, The First FIB Congress*, session 8, vol. 6. Japan Prestressed Concrete Engineering Association, Tokyo, pp. 1–16.

Gjørv, O.E. (2006). The Durability of Concrete Structures in the Marine Environment. In *Durability of Materials and Structures in Building and Civil Engineering*, ed. C.W. Yu and J.W. Bull. Whittles Publishing, Dunbeath, Scotland, pp. 106–127.

Gjørv, O.E., and Kashino, N. (1986). Durability of a 60 Year Old Reinforced Concrete Pier in Oslo Harbour. *Materials Performance*, 25(2), 18–26.

Gjørv, O.E, Sakai, K., and Banthia, N. (eds.) (1998). *Proceedings, Second International Conference on Concrete under Severe Conditions— Environment and Loading*. E & FN Spon, London.

Guofei, L., Årskog, V., and Gjørv, O.E. (2005). *Field Tests with Surface Hydrophobation—Ulsteinvik*, Project Report BML 200503. Department of Structural Engineering, Norwegian University of Science and Technology—NTNU, Trondheim.

Hasselø, J.A. (1997). Ullasundet Bridge—The Life Cycle of a Concrete Structure. In *Proceedings, Seminar on Life Cycle Management of Concrete Structures*. Department of Building Materials, Norwegian University of Science and Technology—NTNU, Trondheim (in Norwegian).

Hasselø, J.A. (2007). Personal information.

Helland, S., Aarstein, R., and Maage, M. (2010). In-Field Performance of North Sea Offshore Platforms with Regard to Chloride Resistance. *Structural Concrete*, 11(1), 15–24.

Hølaas, H. (1992). *Condition of the Concrete Structures at the Statfjord and Gullfaks Oil Fields*, Report OD 92/87. Norwegian Petroleum Directorate, Stavanger (in Norwegian).

Holen Relling, R. (1999). Coastal Concrete Bridges: Moisture State, Chloride Permeability and Aging Effects, Dr. Ing. Thesis 1999:74. Department of Structural Engineering, Norwegian University of Science and Technology—NTNU, Trondheim.

Horrigmoe, G. (2000). Future Needs in Concrete Repair Technology. In *Concrete Technology for a Sustainable Development in the 21st Century*, ed. O.E. Gjørv and K. Sakai. E & FN Spon, London, pp. 332–340.

Kompen, R. (1998). What Can Be Done to Improve the Quality of New Concrete Structures? In *Proceedings, Second International Conference on Concrete Under Severe Conditions—Environment and Loading*, vol. 3, ed. O.E. Gjørv, K. Sakai, and N. Banthia. E & FN Spon, London, pp. 1519–1528.

Knudsen, A., Jensen, F.M., Klinghoffer, O., and Skovsgaard, T. (1998). Cost-Effective Enhancement of Durability of Concrete Structures by Intelligent Use of Stainless Steel Reinforcement. In *Proceedings, Conference on Corrosion and Rehabilitation of Reinforced Concrete Structures*, Florida.

Knudsen, A., and Skovsgaard, T. (1999). Ahead of Its Peers—Inspection of 60 Years Old Concrete Pier in Mexico Reinforced with Stainless Steel. *Concrete Engineering International*.

Lahus, O. (1999). An Analysis of the Condition and Condition Development of Concrete Wharves in Norwegian Fishing Harbours, Dr.Ing. Thesis 1999:23. Department of Building Materials, Norwegian University of Science and Technology—NTNU, Trondheim (in Norwegian).

Lahus, O., Gussiås, A., and Gjørv, O.E. (1998). *Condition, Operation and Maintenance of Norwegian Concrete Harbour Structures*, Report BML 98008. Department of Building Materials, Norwegian University of Science and Technology—NTNU, Trondheim (in Norwegian).

Li, Z.J., Sun, W., Miao, C.W., Sakai, K., Gjørv, O.E, and Banthia, N. (eds.). (2013). *Proceedings, Seventh International Conference on Concrete under Severe Conditions—Environment and Loading*. RILEM, Bagneux.

Malhotra, V.M. (ed.). (1980). *Proceedings, First International Conference on Performance of Concrete in Marine Environment*. ACI SP-65.

Malhotra, V.M. (ed.). (1988). *Proceedings, Second International Conference on Performance on Concrete in Marine Environment*. ACI SP-109.

Malhotra, V.M. (ed.). (1996). *Proceedings, Third International Conference on Performance on Concrete in Marine Environment*. ACI SP-163.

Matta, Z.G. (1993). Deterioration of Concrete Structures in the Arabian Gulf. *Concrete International*, 15, 33–36.

Mehta, P.K. (ed.). (1989). *Proceedings, Ben C. Gerwick Symposium on International Experience with Durability of Concrete in Marine Environment*. Department of Civil Engineering, University of California at Berkeley, Berkeley, California.

Mehta, P.K. (ed.). (1996). *Proceedings, Odd E. Gjørv Symposium on Concrete for Marine Structures*. CANMET/ACI, Ottawa.

Moksnes, J. (1982). Offshore Concrete—Recent Developments in Concrete Mix Design. *Nordisk Betong*, pp. 2–4.

Moksnes, J., and Jakobsen, B. (1985). *High-Strength Concrete Development and Potentials for Platform Design*, OTC Paper 5073. Annual Offshore Technology Conference, Houston, TX, pp. 485–495.

Moksnes, J., and Sandvik, M. (1996). Offshore Concrete in the North Sea—A Review of 25 Years Continuous Development and Practice in Concrete Technology. In *Proceedings, Odd E. Gjørv Symposium on Concrete for Marine Structures*, ed. P.K. Mehta. ACI/CANMET, Ottawa, pp. 1–22.

Nilsson, I. (1991). *Repairs of the Øland Bridge, Experience and Results of the First Six Columns Carried Out in 1990*, Report. NCC, Malmø (in Swedish).

NORDTEST. (1999). *NT Build 492: Concrete, Mortar and Cement Based Repair Materials, Chloride Migration Coefficient from Non-Steady State Migration Experiments*. NORDTEST, Espoo.

NPD. (1976). *Regulations for the Structural Design of Fixed Structures on the Norwegian Continental Shelf*. Norwegian Petroleum Directorate, Stavanger.

NPRA. (1993). *The Helgelands Bridge—Chloride Penetration*, Internal Report. Norwegian Public Road Administration—NPRA, Oslo (in Norwegian).

NTNU. (2005). *Private Archive No. 60: Concrete in Seawater*, University Library, Norwegian University of Science and Technology, Trondheim.

Oh, B.H., Sakai, K., Gjørv, O.E., and Banthia, N. (eds.). (2004). *Proceedings, Fourth International Conference on Concrete under Severe Conditions—Environment and Loading*. Seoul National University and Korea Concrete Institute, Seoul.

Østmoen, T. (1998). Field Tests with Cathodic Protection of the Oseberg A Platform. *Ingeniørnytt*, 34(6), 16–17 (in Norwegian).

Østmoen, T., Liestøl, G., Grefstad, K.A., Sand, B.T., and Farstad, T. (1993). *Chloride Durability of Coastal Concrete Bridges*, Report. Norwegian Public Roads Administration, Oslo (in Norwegian).

Progreso Port Authorities. (2008). Private communication.

Pruckner, F. (2002). Diagnosis and Protection of Corroding Steel in Concrete, PhD Thesis 2002:140. Department of Structural Engineering, Norwegian University of Science and Technology—NTNU, Trondheim.

Pruckner, F., and Gjørv, O.E. (2002). Patch Repair and Macrocell Activity in Concrete Structures. *ACI Materials Journal*, 99(2), 143–148.

Rambøll. (1999). *Pier in Progreso, Mexico—Evaluation of the Stainless Steel Reinforcement*, Inspection Report 990022. Rambøll Consulting Engineers, Copenhagen.

Sakai, K., Banthia, N., and Gjørv, O.E. (eds.). (1995). *Proceedings, First International Conference on Concrete under Severe Conditions—Environment and Loading.* E & FN Spon, London.

Sandvik, M., Haug, A.K., and Erlien, O. (1994). *Chloride Permeability of High-Strength Concrete Platforms in the North Sea*, ACI SP-145, ed. V.M. Malhotra. Detroit, pp. 121–130.

Sandvik, M., and Wick, S.O. (1993). Chloride Penetration into Concrete Platforms in the North Sea. In *Proceedings, Workshop on Chloride Penetration into Concrete Structures*, ed. L.-O. Nilsson. Division of Building Materials, Chalmers University of Technology, Gothenburg.

Sellevold, E.J. (1997). Resistivity and Humidity Measurements of Repaired and Non-Repaired Areas of Gimsøystraumen Bridge. In *Proceedings, Repair of Concrete Structures—From Theory to Practice in a Marine Environment*, ed. A. Blankvoll. Norwegian Road Research Laboratory, Oslo, pp. 283–295.

Sengul, Ő., and Gjørv, O.E. (2007). Chloride Penetration into a 20 Year Old North Sea Concrete Platform. In *Proceedings, Fifth International Conference on Concrete under Severe Conditions—Environment and Loading*, ed. F. Toutlemonde, K. Sakai, O.E. Gjørv, and N. Banthia, vol. 1. Laboratoire Central des Ponts et Chaussées, Paris, pp. 107–116.

Simon, P. (2004). Improved Current Distribution due to a Unique Anode Mesh Placement in a Steel Reinforced Concrete Parking Garage Slab CP System. In *NACE Corrosion 2004*, Paper 04345.

Standard Norway. (1973). *NS 3473: Concrete Structures—Design and Detailing Rules.* Standard Norway, Oslo (in Norwegian).

Standard Norway. (1986). *NS 3420: Specification Texts for Buildings and Construction Works.* Standard Norway, Oslo (in Norwegian).

Standard Norway. (1989). *NS 3473: Concrete Structures—Design and Detailing Rules.* Standard Norway, Oslo (in Norwegian).

Standard Norway. (2003a). *NS-EN 206-1: Concrete—Part 1: Specification, Performance, Production and Conformity*, Amendment prA1:2003 Incorporated. Standard Norway, Oslo (in Norwegian).

Standard Norway. (2003b). NS 3473: *Concrete Structures—Design and Detailing Rules.* Standard Norway, Oslo (in Norwegian).

Stoltzner, E., and Sørensen, B. (1994). Investigation of Chloride Penetration into the Farø Bridges. *Dansk Beton*, 11(1), 16–18 (in Danish).

Stratfull, R.F. (1970). Personal communication.

Thomas, M.D.A., Bremner, T., and Scott, A.C.N. (2011). Actual and Modeled Performance in a Tidal Zone. *Concrete International*, 33(11), 23–28.

Toutlemonde, F., Sakai, K., Gjørv, O.E., and Banthia, N. (eds.). (2007). *Proceedings, Fifth International Conference on Concrete under Severe Conditions—Environment and Loading*. Paris, Laboratoire Central des Ponts et Chauseés, Paris.

Transportation Research Board. (1986). *Strategic Highway Research Program Research Plans*. American Association of State Highway and Transportation Officials.

U.S. Accounting Office. (1979). *Solving Corrosion Problems of Bridge Surfaces Could Save Billions*. Comptroller General of the United States, U.S. Accounting Office PSAD-79-10.

Wood, J.G.M., and Crerar, J. (1997). Tay Road Bridge: Analysis of Chloride Ingress, Variability and Prediction of Long Term Deterioration. *Construction and Building Materials*, 11(4), 249–254.

Yunovich, M., Thompson, N.G., Balvanyos, T., and Lave, L. (2001). *Corrosion Cost and Preventive Strategies in the United States—Highway Bridges*, Appendix D, FHWA-RD-01-156. Office of Infrastructure Research and Development, U.S. Federal Highway Administration.

Chapter 3

Corrosion of embedded steel

3.1 GENERAL

Although there are a number of different deteriorating processes that may cause problems to the durability and performance of concrete structures in severe environments, much international experience has been gained in recent years in developing improved procedures and guidelines for control of both alkali–aggregate reaction and freezing and thawing, although such deteriorating processes may still represent a great challenge in many cases. As shown in Chapter 2, however, proper control of chloride ingress and premature corrosion of embedded steel still appear to be great challenges to both the durability design and the operation of concrete structures in severe environments; occasionally, early chloride ingress may also represent a challenge during concrete construction, as demonstrated in Chapter 2. In the following, therefore, ingress of chlorides and corrosion of embedded steel will be outlined and discussed in more detail.

The generally high ability of concrete to protect embedded steel against corrosion is well known, and this is primarily due to the electrochemical passivation of all embedded steel in the highly alkaline pore solution of the concrete. However, when the passivity partly or completely breaks down due to either concrete carbonation or the presence of chlorides, the corrosion starts, which means that the electrochemical potential of the embedded steel locally becomes more negative and forms anodic areas, while other portions of the steel that have the passive potential intact will act as catchment areas of oxygen and form cathodic areas. If the electrical resistivity of the concrete also is sufficiently low, a rather complex system of galvanic cell activities develops along the embedded steel. In all of these galvanic cells, a flow of current takes place, the amount of which determines the rate of corrosion. Although the size and geometry of the anodic and cathodic areas in these galvanic cells are also important factors, the rate of corrosion is primarily controlled by the electrical resistivity of the concrete and the availability of oxygen for the cathodic process.

For dense, high-quality concrete of proper thickness, carbonation of concrete and carbonation-induced corrosion do normally not represent any problem; for concrete structures in moist environments, it appears from Chapter 2 that concrete carbonation did not represent any problem, even for moderate qualities of concrete.

For concrete structures in severe chloride-containing environments, however, it appears from Chapter 2 that it may just be a question of time before detrimental amounts of chlorides reach embedded steel even through thick covers of high-quality concrete. The high scatter and variability of achieved construction quality also represent a special challenge to the durability and performance of the concrete structures; any weaknesses and deficiencies will soon be revealed, whatever durability specifications and materials have been applied.

3.2 CHLORIDE INGRESS

3.2.1 General

For concrete structures in chloride-containing environments, the ingress of chlorides can take place in different ways. Through uncracked concrete, the ingress mainly takes place by capillary absorption and diffusion. When a relatively dry concrete is exposed to saltwater, the concrete may suck the saltwater relatively fast, and intermittent wetting and drying may successively accumulate high concentrations of salt in the concrete. For concrete structures in marine environments, intermittent exposure to splashing of seawater may also give fluctuating moisture contents in the outer layer of the concrete, as typically shown in Figure 3.1. For many of the concrete structures along the Norwegian coastline, however, constantly high moisture contents in the outer layer of the concrete were observed; for some of the concrete coastal bridges the degree of capillary saturation typically varied from 80 to 90% in the outer 40 to 50 mm of the concrete (Chapter 2). Thus, for the thickness of concrete cover typically specified for concrete structures in severe environments, the moisture content in the concrete cover may be quite high, and hence diffusion appears to be a most dominating transport mechanism for the ingress of chlorides.

Although the ingress of chlorides into concrete has been the subject for extensive research from both a theoretical and an applied point of view, it still appears to be a very complex and difficult issue (Poulsen and Mejlbro, 2006; Tang et al., 2012). Even pure diffusion of chloride ions into concrete is a very complex transport process (Zhang and Gjørv, 1996). Therefore, when Fick's second law of diffusion is often applied for calculating the rates of chloride ingress into concrete, it should be noted that such a calculation

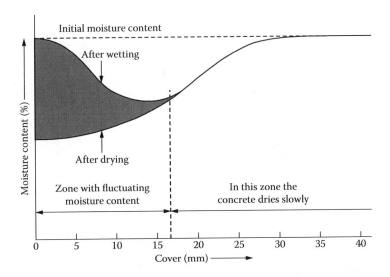

Figure 3.1 Outer layer of the concrete with changing moisture content under splash water conditions along the North Sea coast according to Bakker (1992). (From Bijen, J., *Blast Furnace Slag for Durable Marine Structures*, VNC/BetonPrisma, Hertogenbosch, Netherlands, 1998.)

is based on a number of assumptions and a very rough simplification of the real transport mechanism.

For a proper assessment of the resistance of concrete to chloride ingress, a number of factors have to be considered. Generally, the water/cement ratio of the concrete is a most important controlling factor. In order to get a proper resistance of concrete based on pure portland cements, the water/cement ratio should not exceed a level of 0.40; above this level, the concrete gets a distinctly higher porosity, as shown in Figure 3.2. Although a low water/cement ratio of the concrete is very important, it is well documented in the literature that the selection of a proper cement or binder system may be more important than selecting a low water/cement ratio. Thus, when the water/cement ratio was reduced from 0.50 to 0.40 for a concrete based on a pure portland cement, the chloride diffusivity was reduced by a factor of two to three, while incorporation of various types of supplementary cementitious materials such as blast furnace slag, fly ash, or silica fume at the same water/binder ratio reduced the chloride diffusivity by a factor of up to 20 (Thomas et al., 2011). While a reduced water/cement ratio from 0.45 to 0.35 for a concrete based on pure portland cements may only reduce the chloride diffusivity by a factor of two, a replacement of the portland cement by a proper blast furnace slag cement may reduce the chloride diffusivity by a factor of up to 50 (Bijen, 1998).

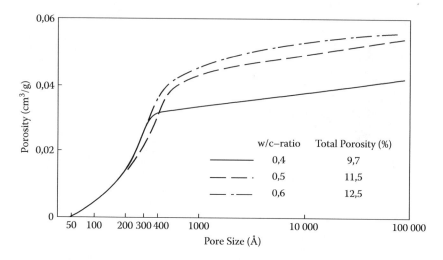

Figure 3.2 Effect of water/cement ratio on the porosity of a concrete based on pure portland cement. (From Gjørv, O. E., and Vennesland, Ø., *Cement and Concrete Research*, 9, 229–238, 1979.)

The beneficial effect of both natural and industrial pozzolanic materials such as condensed silica fume, fly ash, and rice husk ash on the resistance of concrete to chloride ingress is well documented in the literature (Gjørv, 1983; Berry and Malhotra, 1986; Malhotra et al., 1987; FIP, 1998; Malhotra and Ramezanianpour, 1994; Gjørv et al., 1998a). Also, the superior effect of granulated blast furnace slag cements has been documented in the literature for more than 100 years (Bijen, 1998). By combining blast furnace slag cements with various types of pozzolanic material at very low water/binder ratios, very low chloride diffusivities can be obtained, and hence concretes with a very high resistance to chloride ingress can be produced.

3.2.2 Effect of cement type

Figure 3.3 shows the resistance to chloride ingress of four different types of commercial cement produced with the same concrete composition at a water/binder ratio of 0.45. These cements include two blast furnace slag cements of type CEM II/B-S 42.5 R NA with 34% slag (GGBS1) and type CEM III/B 42.5 LH HS (GGBS2) with 70% slag, respectively, one high-performance portland cement of type CEM I 52.5 LA (HPC), and one fly ash cement of type CEM II/A V 42.5 R with 18% fly ash (PFA). The resistance to chloride ingress was determined by use of the rapid chloride migration (RCM) method (NORDTEST, 1999), and all testing was carried out on water-cured concrete specimens at 20°C for periods of up to 180 days.

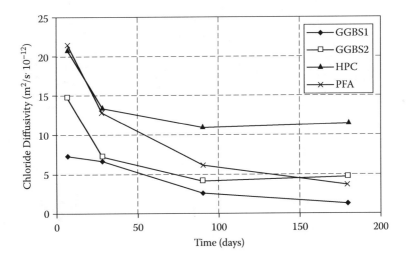

Figure 3.3 Effect of cement type on the resistance of concrete to chloride ingress at a water/binder ratio of 0.45. (From Årskog, V. et al., Effect of Cement Type on the Resistance of Concrete against Chloride Penetration, in *Proceedings, Fifth International Conference on Concrete under Severe Conditions—Environment and Loading*, vol. 1, ed. F. Toutlemonde et al., Laboratoire Central des Ponts et Chauseés, Paris, France, 2007, pp. 367–374.)

In order to compare the same types of cement in a more dense concrete, the same cements were also tested in combination with 10% silica fume by weight of cement at a water/binder ratio of 0.38 (Figure 3.4).

From Figures 3.3 and 3.4 it can be seen that the two slag cements gave a distinctly higher resistance to chloride ingress than the fly ash cement, and a substantially higher resistance than the portland cement. In the more dense concrete, the difference between the different types of cement was smaller than in the more porous type of concrete. However, even in the densest concrete, there was still a distinct difference between the two slag cements and the other cements. Also, both types of slag cement showed a very high early-age resistance to chloride ingress compared to that of the other types of cement. This may be important for an early-age chloride exposure during concrete construction in severe marine environments, as shown in Chapter 2.

In recent years, there has been a rapid trend for using more blended portland cements instead of pure portland cements. Replacement materials such as fly ash and blast furnace slag are also often used as separate additions to the concrete mixture during concrete production. Then, the question is often raised on how much of the portland cement should be replaced in order to obtain a proper resistance to the chloride ingress. While blast furnace slags are hydraulic binders, most types of fly ash are pozzolanic

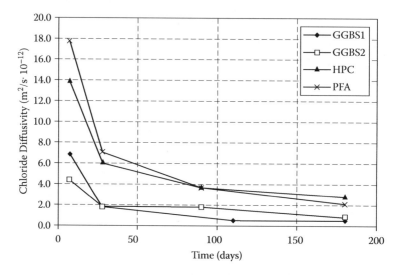

Figure 3.4 Effect of cement type on the resistance of concrete to chloride ingress at a water/binder ratio of 0.38. (From Årskog, V. et al., Effect of Cement Type on the Resistance of Concrete against Chloride Penetration, in *Proceedings, Fifth International Conference on Concrete under Severe Conditions—Environment and Loading*, vol. I, ed. F. Toutlemonde et al., Laboratoire Central des Ponts et Chauseés, Paris, France, 2007, pp. 367–374.)

materials, the main effect of which depends on the amount of $Ca(OH)_2$ available for the pozzolanic reaction. Thus, when the pure portland cement was replaced by more than about 30% low-calcium fly ash, it can be seen from Figure 3.5 that only a very little or no further effect on the resistance to chloride ingress was observed. These results were based on the rapid chloride permeability (RCP) method (ASTM, 2005) on concrete at a water/binder ratio of 0.35 cured in water at 20°C for one year.

In order to establish a proper concrete mixture for a new concrete harbor structure, some preliminary tests with different types of binder systems were carried out on the construction site. These tests included the production of three solid concrete elements, where the high-performance portland cement (CEM I 52.5 LA) was partly replaced by 20, 40, and 60% fly ash, respectively; all concrete mixtures had a water/binder ratio of 0.39. From each test element, concrete cores were removed for the testing of chloride diffusivity at various ages of up to three years of field curing (Tables 3.1 and 3.2). While all results in Table 3.1 were based on RCM testing (NORDTEST, 1999), the results in Table 3.2 were based on parallel testing by use of the immersion method (NORDTEST, 1995). From both tables, however, no further effect on the chloride diffusivity beyond 20% replacement of the portland cement was observed.

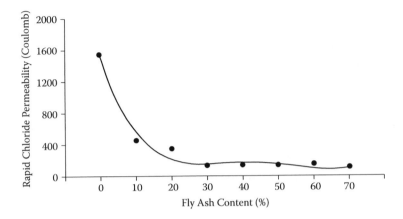

Figure 3.5 Effect of fly ash replacements on the rapid chloride permeability (RCP) after one year of water curing at 20°C. (From Sengul, Ö., Effects of Pozzolanic Materials on Mechanical Properties and Chloride Diffusivity of Concrete, PhD Thesis, Istanbul Technical University, Institute of Science and Technology, Istanbul, Turkey, 2005.)

Table 3.1 Effect of fly ash on the chloride diffusivity (RCM) after up to three years of field curing on the construction site[a]

Concrete	I year	2 years	3 years
20% FA	1.4	1.1	0.66
	0.2	0.2	0.27
40% FA	1.4	1.2	1.13
	0.1	0.1	0.38
60% FA	1.7	1.5	1.07
	0.1	0.1	0.22

Source: Årskog, V., and Gjørv, O. E., *Container Terminal Sjursøya—Low Heat Concrete—Three-Year Durability*, Report BML200901, Department of Structural Engineering, Norwegian University of Science and Technology—NTNU, Trondheim, Norway (in Norwegian), 2009.

[a] Mean values and standard deviation ($\times 10^{-12}$ m²/s).

Although portland cements can be replaced by larger amounts of blast furnace slag compared to that of fly ash, for slag additions there also appears to be an upper limit above which the observed effect is very small. Thus, in Figure 3.6 the portland cement (CEM I 42.5 R) was partly replaced by 40, 60, and 80% blast furnace slag with a Blaine fineness of 5,000 cm²/g. Based on concrete with a water/binder ratio of 0.40 and water curing at 20°C of

Table 3.2 Effect of fly ash on the chloride diffusivity
(immersion method) after two years of field
curing on the construction site[a]

Chloride diffusivity $(\times 10^{-12}\ m^2/s)$	20% FA	40% FA	60% FA
	0.48	0.44	0.50
	0.16	0.05	0.13

Source: Årskog, V., and Gjørv, O. E., *Container Terminal Sjursøya—Low Heat Concrete—Three-Year Durability*, Report BML200901, Department of Structural Engineering, Norwegian University of Science and Technology—NTNU, Trondheim, Norway (in Norwegian), 2009.

[a] Mean values and standard deviation.

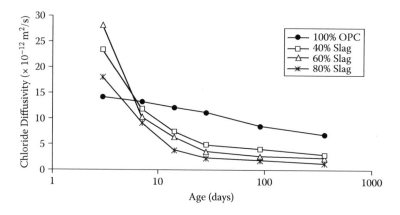

Figure 3.6 Effect of blast furnace slag on the chloride diffusivity of concrete (RCM). (From Sengul, Ö., and Gjørv, O. E., Effect of Blast Furnace Slag for Increased Concrete Sustainability, in *Proceedings, International Symposium on Sustainability in the Cement and Concrete Industry*, ed. S. Jacobsen et al., Norwegian Concrete Association, Oslo, Norway, 2007, pp. 248–256.)

up to one year, the resistance of the concrete to chloride ingress was tested by use of the RCM method. After 28 days, it can be seen that increasing amounts of slag successively reduced the chloride diffusivity from 11.2 to 4.9, 3.6, and 2.3 × 10^{-12} m^2/s, respectively, while after one year, the diffusivity of the slag concretes varied from 3.0 to 1.2 × 10^{-12} m^2 compared to 7.0 × 10^{-12} m^2/s for that of the pure portland cement. In parallel, diffusion testing by use of the immersion method also showed a similar effect of the increased replacements of the portland cement by slag (Figure 3.7). After 28 days of water curing and a further 35 days of immersion in the salt solution, the chloride diffusivity was reduced from 12.8 × 10^{-12} m^2/s for the pure portland cement to 4.0, 3.1, and 3.2 × 10^{-12} m^2/s for the 40, 60, and 80% slag contents, respectively. All these test results are in general agreement with other results reported in the literature (Bijen, 1998). Thus, from

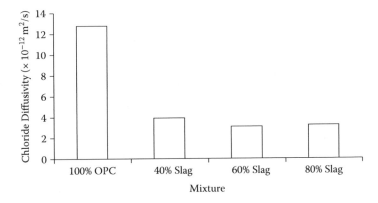

Figure 3.7 Effect of blast furnace slag on the chloride diffusivity of concrete (immersion). (From Sengul,Ö., *Effects of Pozzolanic Materials on Mechanical Properties and Chloride Diffusivity of Concrete*, PhD Thesis, Istanbul Technical University, Institute of Science and Technology, Istanbul, Turkey, 2005.)

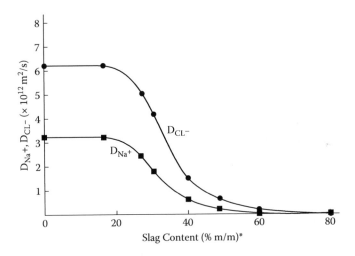

Figure 3.8 Effect of blast furnace slag on the diffusivity of pure cement paste at a water/binder ratio of 0.60 according to Brodersen (1982). (From Bijen, J., *Blast Furnace Slag for Durable Marine Structures*, VNC/BetonPrisma, Hertogenbosch, Netherlands, 1998.)

Figure 3.8 it can be seen that there was hardly any effect on the chloride diffusivity for slag contents of less than about 25%, while for slag contents of 25 to 50%, there was a large drop in diffusivity, beyond which there was still a decrease, but only to a smaller extent.

For many of the cement replacement materials, further beneficial effects on the resistance to chloride ingress can be obtained by increased fineness of the replacement material. Thus, by grinding the blast furnace slag up to a Blaine fineness of 16,000 cm²/g, an extremely high resistance to chloride ingress was observed (Gjørv et al., 1998c). Based on the steady-state migration method (NORDTEST, 1989), ultimate chloride diffusivities in the range of 0.04 to 0.08 × 10⁻¹² m²/s were obtained. Also, by replacing the portland cement with 30% blast furnace slag of a 8,700 cm²/g Blaine fineness and 10% silica fume, both a very high early-age and ultimate resistance to chloride ingress were observed (Teng and Gjørv, 2013). Thus, after 3 and 28 days of curing, the chloride diffusivities were 1.7 and 0.1 × 10⁻¹² m²/s (RCM), respectively, while after 90 days, the chloride diffusivity had reached a value of 0.01 × 10⁻¹² m²/s.

In the literature, the beneficial effect of C_3A in pure portland cements for a chemical binding of the penetrating chlorides is often referred to. In Figure 3.9, where the chloride ingress of the two different portland cements with 0 and 8.6% C_3A, respectively, is shown, no beneficial effect of C_3A can be seen. Again, the superior resistance of the 80% blast furnace slag cement to chloride ingress compared to that of the two types of portland cements, the 30% slag cement and the 26% trass cement, is demonstrated. These

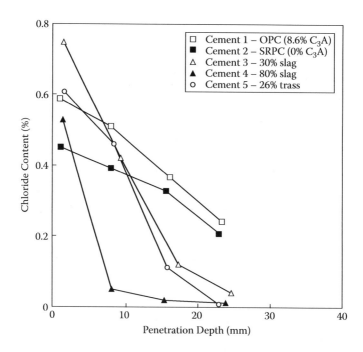

Figure 3.9 Effect of cement type on chloride ingress (by weight of cement) into concrete exposed to fresh circulating seawater. (From Gjørv, O. E., and Vennesland, Ø., *Cement and Concrete Research*, 9, 229–238, 1979.)

results were based on field tests of mortar with a water/binder ratio of 0.50 submerged in fresh circulating seawater at a temperature of about 7°C. According to Mehta (1977), chemical binding of penetrating chlorides cannot be expected unless the C_3A content is much higher than 8%. For a concrete produced with a 16% C_3A type of portland cement at a water/cement ratio of 0.34, however, a chloride ingress of up to 80 mm was observed after about 100 years of exposure to seawater in a Japanese concrete harbor structure (Gjørv et al., 1998b).

According to Sluijter (1973) the binding capacity of penetrating chlorides in a hydrated cement paste is primarily a matter of physical adsorption to the surface of the CSH-gel rather than chemical binding. For both portland cements blended with pozzolanic materials and blast furnace slag cements, a substantially higher formation of CSH-gel with a higher amount of small gel pores (<30 nm) and a smaller amount of large capillary pores than those of pure portland cements is achieved. For the 80% slag cement shown in Figure 3.9, as much as about 80% of the total porosity was made up by pores smaller than 200 Å, while for the two portland cements, the corresponding number was only about 30% (Figure 3.10).

Investigations further indicate that the chloride binding capacity of the slag cements may also be due to the higher aluminate levels in the slag forming higher amounts of Friedel's salt (Dihr et al., 1996). The substantially

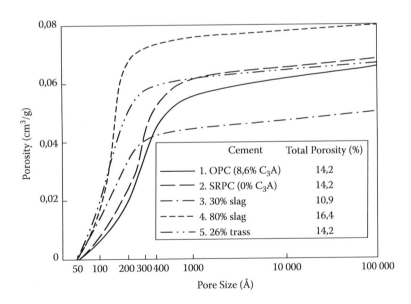

Figure 3.10 Effect of cement type on pore size distribution and porosity of the concrete. (From Gjørv, O. E., and Vennesland, Ø., *Cement and Concrete Research*, 9, 229–238, 1979.)

smaller amount of free lime in the pore solution may also be beneficial for a low chloride diffusivity.

However, a reduced amount of free lime reduces the alkalinity of the pore solution, which may also reduce the critical level of the chloride concentration for breaking the passivity of embedded steel. For very dense concrete, however, such a reduced critical level of chloride concentration does not necessarily represent any practical durability problem. The slag also very much increases the electrical resistivity of the concrete in such a way that an ohmic control of the further corrosion process is typically obtained.

As already pointed out, even a pure diffusion of chloride ions into concrete from an external salt solution is a very complex transport process. As part of this process, the chemical composition of the external salt solution is also very important for the resulting ingress of chlorides (Theissing et al., 1975). This effect is clearly demonstrated in Figure 3.11, where a deeper

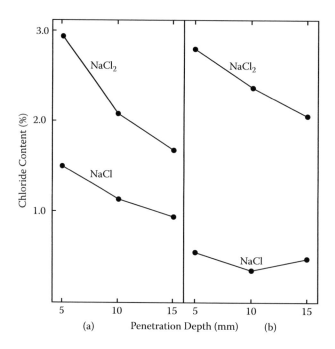

Figure 3.11 Chloride ingress (by weight of cement) into cement paste from two different types of salt solution of the same chloride concentration: (a) after 2 days of hydration and (b) after 40 days of hydration before exposure. (From Trætteberg, A., *The Mechanism of Chloride Penetration into Concrete*, SINTEF Report STF65 A77070, Trondheim, Norway [in Norwegian], 1977; Gjørv, O. E., and Vennesland, Ø., Evaluation and Control of Steel Corrosion in Offshore Concrete Structures, in *Proceedings, the Katharine and Bryant Mather International Conference*, vol. 2, ed. J. Scanlon, ACI SP-I, 1987, pp. 1575–1602.)

chloride ingress from a salt solution based on $CaCl_2$ compared to that of NaCl can be seen, the chloride concentration in the external salt solutions being the same. Therefore, not only the porosity of the concrete and its capacity for chloride binding, but also the total ion exchange capacity of the whole system is very important for the resulting chloride ingress. Thus, as demonstrated in Figure 3.11, the use of de-icing salts based on $CaCl_2$ represents a more severe exposure for chloride ingress into concrete than that of NaCl from the seawater.

3.2.3 Effect of temperature

As discussed in Chapter 2, the temperature is also an important controlling factor for the rate of chloride ingress, the effect of which needs special attention for a proper durability design. Since the temperature may also affect the rate of hydration of the various types of cement and binder system, this effect of the temperature may also need special attention in the durability design. As was demonstrated in Chapter 2, the risk for early-age chloride exposure before the concrete has gained sufficient maturity and density may occasionally be very high during concrete construction; in particular, this may be a challenge during periods for concrete construction at low temperatures.

In order to test the effect of low curing temperature on the early-age resistance to chloride ingress, concrete produced with four different commercial cements was tested at curing temperatures of 5, 12, and 20°C, respectively (Figures 3.12–3.14). These cements included one blast furnace slag cement of type CEM III/B 42.5 LH HS (GGBS2) with 70% slag, one high-performance portland cement of type CEM I 52.5 LA (HPC), one ordinary portland cement of type CEM I 42.5 R (OPC), and one fly ash cement of type CEM II/A V 42.5 R with 18% fly ash (PFA). A concrete composition with a water/binder ratio of 0.45 was used, and the resistance to chloride ingress was tested by use of the RCM method. All testing was carried out on water-cured concrete specimens for periods of up to 90 days.

At all curing temperatures, it can be seen from Figures 3.12–3.14 that the 70% slag cement (GGBS2) gave a substantially higher resistance to the chloride ingress than both the fly ash cement (PFA) and the two portland cements (HPC and OPC). Thus at 5°C, the 28-day chloride diffusivity for the slag cement was 7.9×10^{-12} m²/s compared to 17.4×10^{-12} m²/s for the fly ash cement (PFA) and 19.3 and 20.3×10^{-12} m²/s for the two pure portland cements (HPC and OPC), respectively. After 90 days of curing, the corresponding values were 4.1, 17.2, 17.6, and 14.5×10^{-12} m²/s, respectively.

Regardless of curing temperature, the results in Figures 3.12–3.14 demonstrate how the slag cement gave the highest resistance, while the two portland cements gave the lowest resistance to chloride ingress for curing periods of up to 90 days. For severe marine environments with low

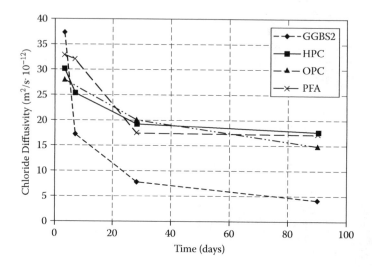

Figure 3.12 Effect of cement type on the resistance of concrete against chloride pene-
tration at a curing temperature of 5°C. (From Liu, G., and Gjørv, O. E., Early
Age Resistance of Concrete against Chloride Penetration, in *Proceedings,
Fourth International Conference on Concrete under Severe Conditions—
Environment and Loading*, vol. 1, ed. B. H. Oh et al., Seoul National University
and Korea Concrete Institute, Seoul, South Korea, 2004, pp. 165–172.)

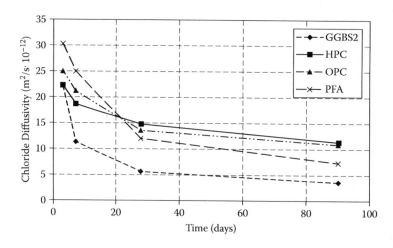

Figure 3.13 Effect of cement type on the resistance of concrete against chloride pene-
tration at a curing temperature of 12°C. (From Liu, G., and Gjørv, O. E., Early
Age Resistance of Concrete against Chloride Penetration, in *Proceedings,
Fourth International Conference on Concrete under Severe Conditions—
Environment and Loading*, vol. 1, ed. B. H. Oh et al., Seoul National University
and Korea Concrete Institute, Seoul, South Korea, 2004, pp. 165–172.)

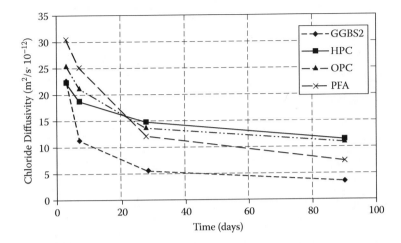

Figure 3.14 Effect of cement type on the resistance of concrete against chloride pene-
tration at a curing temperature of 20°C. (From Liu, G., and Gjørv, O. E., *Early
Age Resistance of Concrete against Chloride Penetration*, in *Proceedings,
Fourth International Conference on Concrete under Severe Conditions—
Environment and Loading*, vol. I, ed. B. H. Oh et al., Seoul National University
and Korea Concrete Institute, Seoul, South Korea, 2004, pp. 165–172.)

curing temperatures, the above results clearly demonstrate that concrete
structures produced with portland cements or fly ash cements are much
more vulnerable to early-age chloride exposure than structures produced
with a binder system based on blast furnace slag.

3.3 PASSIVITY OF EMBEDDED STEEL

The pore solution of concrete based on portland cements normally attains
an alkalinity level in excess of pH 13. In the presence of oxygen, this alkaline solution forms a thin oxide film on the steel surface that very efficiently
protects all embedded steel from corrosion. However, the integrity and
protective quality of this film depends on a number of factors, such as oxygen availability and alkalinity of the pore solution, and the lower the
oxygen availability and the lower the alkalinity, the thinner the protective
film and the lower the protective quality. Experience indicates that the pH
of the pore solution should never drop below a level of approximately 11.5
in order to maintain a proper electrochemical protection of the embedded
steel (Shalon and Raphael, 1959). As soon as the pH drops to approximately 9.0, however, the protective oxide film is completely dissolved and
broken down.

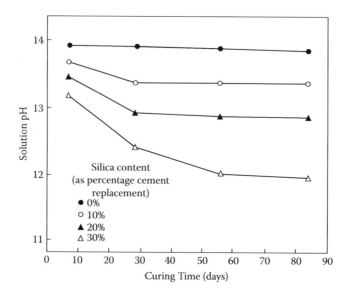

Figure 3.15 Effect of increased additions of silica fume on the basicity of concrete based on portland cement. (From Page, C. L., and Vennesland, Ø., *Materials and Structures*, RILEM, 16(91), 19–25, 1983.)

For concrete based on portland cements, the high alkalinity is due to small quantities of the readily soluble NaOH and KOH. The cement paste also contains a large proportion of the less soluble $Ca(OH)_2$, which buffers the system in such a way that the pH never drops below a level of approximately 12.5. For granulated blast furnace slag cements and portland cements blended with pozzolans such as fly ash or condensed silica fume, however, certain amounts of the $Ca(OH)_2$ are bound up to form new CSH, and hence the reserve basicity is correspondingly reduced. This is demonstrated in Figure 3.15, where increased additions of silica fume successively reduced the basicity of the concrete.

Even for cements with the highest reserve basicity, the alkalinity may still be reduced either by leaching of the alkaline substances with water or by neutralization after carbonation with CO_2. The pore solution in carbonated concrete only has a pH of approximately 8.5.

As previously discussed, carbonation of a dense, high-quality concrete is not considered to represent any practical problem. However, the protective oxide film can easily be destroyed by the presence of chlorides in the concrete, and the thinner the oxide film, the less the amounts of chlorides needed in order to destroy the protective film. It is well known that even very small chloride contents in the pore solution may destroy the passivity of the steel. However, only a small part of the total amount of the chlorides in concrete is dissolved in the pore solution. Some of the chlorides are chemically bound, and some are physically bound, while the rest exist in the form of free chlorides dissolved in the pore solution. It is only these free chlorides in the pore solution that can destroy the protective film and thus start corrosion. Since a very complex equilibrium between the different forms of chlorides in the concrete exists, however, the amount of free chlorides in the pore solution of a given concrete depends on both the degree of water saturation and the temperature of the concrete.

Different types of binder system may also significantly affect both the pore solution alkalinity and the amount of chemically and physically bound chlorides in the concrete. Whether the chlorides have been mixed in already during concrete production or have penetrated the concrete later on may also affect the above complex relationship. If the mixed-in chloride content has been too high already from the beginning, the protective film on the embedded steel may never have been formed.

The threshold concentration of chlorides required to destroy the passivity of embedded steel has been the subject for numerous investigations (Bertolini et al., 2004; Angst, 2011). A number of different measurement techniques have been applied, and a number of different threshold values reported, typically varying from 0.02 to 3.04% of total chloride content by weight of cement. There are a number of factors affecting the chloride threshold, such as pH of the pore solution, potential of the steel, and the local conditions along the interface between the concrete and the embedded steel (Glass and Buenfeld, 1997, 2000). As a result, it is not possible to express any unique chloride threshold value for corrosion of embedded steel in concrete. However, only very small amounts of chlorides are needed for breaking the passivity, and as soon as the passivity is broken and the corrosion starts, the rate of corrosion is then controlled by a number of other factors.

3.4 CORROSION RATE

3.4.1 General

For ongoing corrosion in concrete structures exposed to the atmosphere, experience indicates that the corrosion rate may vary from several tens of μm/year to localized values of up to 1 mm/year, depending on the moisture conditions and the chloride contents in the concrete (Bertolini et al., 2004). Different temperatures may also affect the corrosion rate very differently (Andrade and Alonso, 1995; Østvik, 2005). Only for concrete structures permanently submerged in seawater, it was shown in Chapter 2 that the corrosion rate was so low that the observed damage was negligible even after a service period of more than 60 years (Gjørv and Kashino, 1986). In Chapter 2, it was also shown how ongoing corrosion in certain parts of the rebar system had effectively protected other parts of the rebar system cathodically. Therefore, since both the geometrical shape of the structure and the local environmental conditions are very important factors for the rate of corrosion, it appears very difficult to mathematically predict the long-term effect of ongoing corrosion on the total load carrying capacity of a concrete structure. For the corrosion to develop into a serious deteriorating process, however, the electrical resistivity of the concrete is also a most important controlling factor.

3.4.2 Electrical resistivity

Since the electrical current flowing in all the galvanic cells along the embedded steel is transported by charged ions through the concrete, the electrical resistivity of the concrete depends on the permeability of the concrete, the amount of pore solution, and the ion concentration of the pore solution. Thus, by decreasing the water/cement ratio from 0.7 to 0.5, it can be seen from Figure 3.16 that the electrical resistivity increased by a factor of more than two for mortar, while for concrete of the same water/cement ratio,

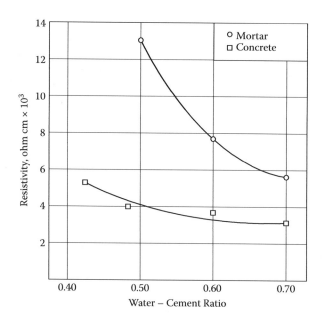

Figure 3.16 Effect of water/cement ratio on the electrical resistivity of concrete. (From Gjørv O. E. et al., Electrical Resistivity of Concrete in the Ocean, in *Proceedings, Ninth Annual Offshore Technology Conference*, OTC Paper 2803, Houston, Texas, 1977, pp. 581–588.)

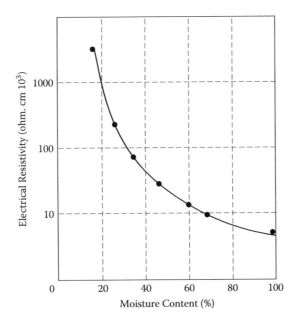

Figure 3.17 Effect of moisture conditions on the electrical resistivity of concrete. (From Gjørv O. E. et al., Electrical Resistivity of Concrete in the Ocean, in *Proceedings, Ninth Annual Offshore Technology Conference*, OTC Paper **2803**, Houston, Texas, 1977, pp. 581–588.)

there was only a small effect. These results clearly demonstrate the effect of permeability on the electrical resistivity. When the degree of water saturation of the concrete was successively decreased from 100% to somewhat less than 20% relative humidity (RH), the electrical resistivity increased from approximately 7×10^3 to approximately 6×10^6 ohm cm, as shown in Figure 3.17. These results clearly demonstrate the very important effect of the moisture conditions of the concrete, and thus the amount of pore solution available for the transport of charged ions. When the portland cement in the concrete was successively replaced by condensed silica fume, both the permeability and the ion concentration of the pore solution were significantly affected, the great effect of which is demonstrated in Figure 3.18.

If the electrical resistivity of the concrete becomes sufficiently high, a very small or negligible rate of corrosion may take place. From the extensive condition assessments of the San Mateo–Hayward Bridge, a threshold value of $50–70 \times 10^3$ ohm cm for the concrete was reported, beyond which only a very small corrosion rate was observed (Chapter 2). Even for the combination of broken passivity and low electrical resistivity, however, the rate of corrosion may still be very low or negligible, depending on the availability of oxygen.

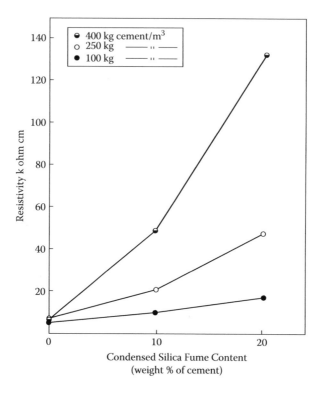

Figure 3.18 Effect of increased additions of silica fume on the electrical resistivity of concrete. (From Vennesland, Ø., and Gjørv, O. E., Silica Concrete— Protection against Corrosion of Embedded Steel, in *Proceedings, CANMET/ ACI International Conference on Fly Ash, Silica Fume, and Natural Pozzolans in Concrete*, vol. II, ACI SP-79, ed. V. M. Malhotra, 1983, pp. 719–729.)

3.4.3 Oxygen availability

The availability of oxygen depends on several factors. While the concentration of oxygen in the atmosphere is approximately 210 ml/l, the maximum concentration of oxygen in water available for corrosion in submerged concrete structures is only 5–10 ml/l. The rate of oxygen diffusion through the concrete cover also depends on whether the oxygen is in a gaseous state or dissolved in water. Although both the permeability and thickness of the concrete cover affect the oxygen availability, it can be seen from Figure 3.19 that the degree of water saturation of the concrete is a most dominating factor. For oxygen to take part in the electrochemical cathode reaction on embedded steel, it must be in a dissolved state. For concrete submerged in water, Figure 3.20 demonstrates that the thickness of the concrete cover may only have a minor effect on the oxygen availability. Thus, for a concrete with a water/cement ratio of 0.40, a reduced concrete

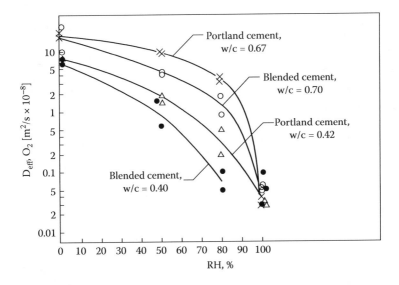

Figure 3.19 Effect of water saturation on the rate of oxygen diffusion. (From Tuutti, K., *Corrosion of Steel in Concrete*, Report 4-82, Cement and Concrete Research Institute, Stockholm, Sweden, 1982.)

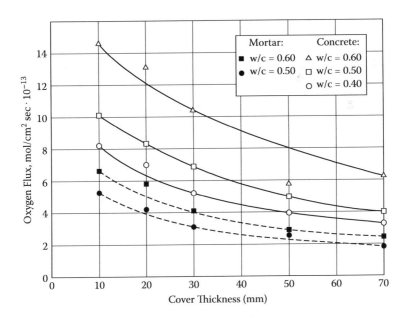

Figure 3.20 Effect of concrete cover on the rate of oxygen diffusion through concrete submerged in water. (From Gjørv, O. E. et al., *Materials Performance*, 25(12), 39–44, 1986.)

cover from 70 to 10 mm only reduced the flux of oxygen by a factor of approximately 2.6. These results indicate that there is a transition phase between the concrete and the embedded steel that acts as a barrier to the oxygen, and this may explain why the thickness of the concrete cover is not so important for the availability of dissolved oxygen.

While corrosion of embedded steel may not represent any practical problem in submerged concrete structures due to lack of dissolved oxygen, macro cell corrosion may still develop due to availability of oxygen from the inside of submerged hollow concrete structures (Bertolini et al., 2004). For concrete structures above water, however, the oxygen is so plentifully available that it does not represent any limiting factor for high corrosion rates to develop.

3.5 CRACKS

For concrete structures with cracks in the concrete cover, the electrolytic conditions for start of corrosion may be significantly affected. For cracked concrete, it is reasonable to assume that increased ingress and availability of corrosive substances such as chlorides, water, and oxygen will give an increased probability of local corrosion. Based on detailed procedures for calculation of crack widths, therefore, most codes and recommendations specify upper limitations for characteristic crack widths of typically 0.4 mm for non-aggressive environments and 0.3 mm for more aggressive environments.

In spite of a large number of experiments reported in the literature, however, it is not possible to come up with a simple relationship between crack width and probability of corrosion for a given concrete structure in a given environment. Extensive research has revealed that a number of factors affect the possibility for corrosion, and both the possible mechanisms and the practical consequences have been the subject for extensive discussions (Gjørv, 1989). As shown in Figure 3.21, the geometry of the cracks also may be very complex, and different effects of the cracks running parallel or perpendicular to the steel bars are observed. Cracks in different types of environment and cracks due to different loading conditions, such as static or dynamic loading, may also affect the observed corrosion differently.

Most of the experimental investigations on the effect of cracks reported in the literature have been carried out on concrete exposed to various types of atmospheric environments. In most of these investigations, it has not been possible to establish a simple relationship between crack width and development of corrosion. Very often, a certain effect of the crack width could be observed at an early stage of exposure, while later on, the observed effect could be very small or almost negligible.

For concrete structures continuously submerged in seawater, the effect of cracks cannot be evaluated without also taking into account the galvanic

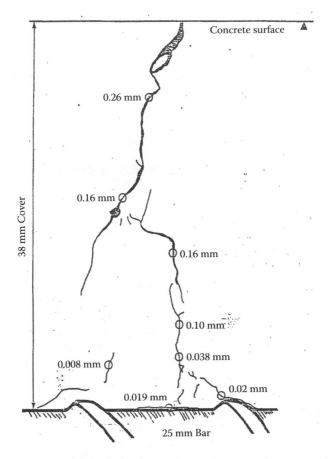

Figure 3.21 The geometry of cracks may be very complex. (From Beeby, A. W., *Cracking and Corrosion*, Report 2/11, UK Research Program on Concrete in the Oceans, 1977.)

coupling between the exposed steel in the crack and the larger portions of the embedded steel system adjacent to the crack (Gjørv, 1977). Thus, results obtained on the basis of small concrete elements cannot necessarily be extrapolated to large submerged concrete structures.

By simulating the galvanic coupling and the corrosion mechanism that may take place in large submerged concrete structures, laboratory investigations revealed that the observed corrosion in the cracked concrete was significantly less than expected (Vennesland and Gjørv, 1981). This was also true for similar tests carried out under dynamic loading conditions (Espelid and Nilsen, 1988).

Depending on the environmental conditions, it appears that the rate of corrosion is reduced over time by a clogging up of the crack by both corrosion products and other reaction products. In particular, this effect appears to be effective for cracked concrete under submerged conditions in seawater. Smaller areas of the freely exposed steel in the crack may also be cathodically protected by the adjacent embedded steel. However, if the areas of the freely exposed steel in galvanic coupling with embedded steel in submerged concrete structures become too high, the rate of corrosion may be significant due to extremely high cathode-to-anode area ratios (Gjørv, 1977).

3.6 GALVANIC COUPLING BETWEEN FREELY EXPOSED AND EMBEDDED STEEL

For large submerged concrete structures such as that used for oil and gas exploration offshore, there are a variety of external steel components, such as skirts, pipes, supports, and fixtures, that are in electrical contact with the embedded reinforcing steel system. As the exposed external steel will then be anodic against the embedded steel, which will be cathodic, special problems both to the corrosion rate and to the corrosion protection of such external steel components may arise (Gjørv, 1977). The rate of corrosion will then primarily be controlled by the cathode-to-anode area ratio and by the cathode efficiency of the embedded steel. Hence, for cathodic protection of such freely exposed steel, the current demand will also be controlled by both the area of the cathode and the cathode efficiency of all the embedded steel system.

For a proper assessment of the cathode-to-anode area ratio, however, both the structural design and the internal electrical continuity within the rebar system must be evaluated on an individual basis. Measurements on large offshore concrete platforms, however, indicate that such heavily reinforced structures have a very good electrical continuity within more or less the whole embedded rebar system. Hence, the cathode-to-anode area ratio for a small area of external steel components may be extremely high relative to the area of all the embedded steel. The cathode efficiency, which depends on the rate of oxygen diffusion through the concrete cover to the embedded steel, is also an important factor. Based on laboratory investigations on submerged concrete of the same quality as that typically used for concrete platforms in the North Sea, oxygen availability with flux values of up to 0.5×10^{13} mol O_2/s and cm^2 were observed (Gjørv et al., 1986). In the field, however, experience from existing concrete platforms in the North Sea has also shown that both marine growth and biological activities on the concrete surface successively reduce the oxygen availability. Only at an early stage of exposure, field investigations revealed a high current drain to the embedded steel, and hence high consumption rates of the sacrificial anodes, but later on this current drain appears to be reduced (Espelid, 1996).

3.7 STRUCTURAL DESIGN

As was typically observed and discussed for all the concrete harbor structures in Chapter 2, all structures with a flat type of deck showed a much better durability and performance than structures with a beam and slab type of deck (Figure 2.13). For a beam and slab type of deck, the more exposed deck beams will always absorb and accumulate more chlorides, and hence develop anodic areas, while the less exposed parts, such as the slab sections in between, will act as catchment areas for oxygen, and hence form cathodic areas. Therefore, the more exposed parts of the deck, such as beams and girders, will always be more vulnerable to steel corrosion than the rest of the concrete structure. In Chapter 2, it was also shown how the corroding beams and girders in the jetty of Oslo Harbor (1922) had effectively functioned as sacrificial anodes, and thus cathodically protected the slab sections in between during a period of up to 60 years. Such examples clearly demonstrate how geometrical shape and structural design may distinctly affect the durability and performance of concrete structures in severe environments.

From a durability point of view, there may also be a good strategy to base the structural design on prefabricated structural elements where possible. Such prefabricated elements may be produced in a more protected and controlled way during construction, as discussed later in Chapter 5.

REFERENCES

Andrade, C., and Alonso, C. (1995). Corrosion Rate Monitoring in the Laboratory and On-Site. In *Construction and Building Materials*. Elsevier Science, London, pp. 315–328.

Angst, U. (2011). Chloride Induced Reinforcement Corrosion in Concrete, Ph.D. Thesis 2011:113. Department of Structural Engineering, Norwegian University of Science and Technology—NTNU, Trondheim.

Årskog, V., Ferreira, M., Liu, G., and Gjørv, O.E. (2007). Effect of Cement Type on the Resistance of Concrete against Chloride Penetration. In *Proceedings, Fifth International Conference on Concrete under Severe Conditions—Environment and Loading*, vol. 1, ed. F. Toutlemont, K. Sakai, O.E. Gjørv, and N. Banthia. Laboratoire Central des Ponts et Chausées, Paris, pp. 367–374.

Årskog, V., and Gjørv, O.E. (2009). *Container Terminal Sjursøya—Low Heat Concrete—Three-Year Durability*, Report BML200901. Department of Structural Engineering, Norwegian University of Science and Technology—NTNU, Trondheim (in Norwegian).

ASTM. (2005). *ASTM C 1202-05: Standard Test Method for Electrical Indication of Concrete's Ability to Resist Chloride Ion Penetration*. ASTM International, West Conshohocken, PA.

Bakker, R.F.M. (1992). *The Critical Chloride Content in Reinforced Concrete*, CUR-Report. Gouda.

Beeby, A.W. (1977). *Cracking and Corrosion*, Report 2/11. UK Research Program on Concrete in the Oceans.

Berry, E.E., and Malhotra, V.M. (1986). *Fly Ash in Concrete*, CANMET SP85. Ottawa.

Bertolini, L., Elsener, B., Pediferri, P., and Polder, R. (2004). *Corrosion of Steel in Concrete*. Wiley-VCH, Weinheim.

Bijen, J. (1998). *Blast Furnace Slag for Durable Marine Structures*. VNC/BetonPrisma, Hertogenbosch.

Brodersen, H.A. (1982). The Dependence of Transport of Various Ions in Concrete from Structure and Composition of the Paste, Dissertation RWTH. Aachen (in German).

Dihr, R.K., El-Mohr, M.A.K., and Dyer, T.D. (1996). Chloride Binding in GGBS Concrete. *Cement and Concrete Research*, 26(12), 1767–1773.

Espelid, B. (1996). *Cathodic Protection of Concrete Structures—Current Drain to Reinforcement*, Technical Report BGN R795253, Det Norske Veritas—DNV, Bergen.

Espelid, B., and Nilsen, N. (1988). *A Field Study of the Corrosion Behavior on Dynamically Loaded Marine Concrete Structures*, ACI SP-109, ed. V.M. Malhotra, pp. 85–104.

FIP. (1998). *Condensed Silica Fume in Concrete—State of the Art Report*. Féderation Internationale de la Précontrainte—FIP, Thomas Telford, London.

Gjørv, O.E. (1977). Corrosion of Steel in Offshore Concrete Platforms. In *Proceedings, Conference on the Ocean—Our Future*. Norwegian Institute of Technology—NTH, Trondheim, pp. 390–401 (in Norwegian).

Gjørv, O.E. (1983). Durability of Concrete Containing Condensed Silica Fume. In *Proceedings, CANMET/ACI International Conference on Fly Ash Silica Fume and Natural Pozzolans in Concrete*, vol. II, ACI SP-79, ed. V.M. Malhotra, pp. 695–708.

Gjørv, O.E. (1989). Mechanisms of Corrosion of Steel in Concrete Structures. In *Proceedings, International Conference on Evaluation of Materials Performance in Severe Environments*, vol. 2. Iron and Steel Institute of Japan, Tokyo, pp. 565–578.

Gjørv, O.E., and Kashino, N. (1986). Durability of a 60 Year Old Reinforced Concrete Pier in Oslo Harbour. *Materials Performance*, 25(2), 18–26.

Gjørv, O.E., Ngo, M.H., and Mehta, P.K. (1998a). Effect of Rice Husk Ash on the Resistance of Concrete against Chloride Penetration. In *Proceedings, Second International Conference on Concrete under Severe Conditions—Environment and Loading*, vol. 3, ed. O.E. Gjørv, K. Sakai, and N. Banthia. E & FN Spon, London, pp. 1819–1826.

Gjørv, O.E., Ngo, M.H., Sakai, K., and Watanabe, H. (1998c). Resistance against Chloride Penetration of Low-Heat High-Strength Concrete. In *Proceedings, Second International Conference on Concrete under Severe Conditions—Environment and Loading*, vol. 3, ed. O.E. Gjørv, K. Sakai, and N. Banthia. E & FN SPON, London, pp. 1827–1833.

Gjørv, O.E., Tong, L., Sakai, K., and Shimizu, T. (1998b). Chloride Penetration into Concrete after 100 Years of Exposure to Seawater. In *Proceedings, Second International Conference on Concrete under Severe Conditions—Environment and Loading*, vol. 1, ed. O.E. Gjørv, K. Sakai, and N. Banthia. E & FN Spon, London, pp. 198–206.

Gjørv, O.E., and Vennesland, Ø. (1979). Diffusion of Chloride Ions from Seawater into Concrete. *Cement and Concrete Research*, 9, 229–238.

Gjørv, O.E., and Vennesland, Ø. (1987). Evaluation and Control of Steel Corrosion in Offshore Concrete Structures. In *Proceedings, the Katharine and Bryant Mather International Conference*, vol. 2, ed. J. Scanlon, ACI SP-1, pp. 1575–1602.

Gjørv, O.E., Vennesland, Ø., and El-Busaidy A.H.S. (1977). Electrical Resistivity of Concrete in the Ocean. In *Proceedings, Ninth Annual Offshore Technology Conference*, OTC Paper 2803, Houston, pp. 581–588.

Gjørv, O.E., Vennesland, Ø., and El-Busaidy A.H.S. (1986). Diffusion of Dissolved Oxygen through Concrete. *Materials Performance*, 25(12), 39–44.

Glass, G.K., and Buenfeld, N.R. (1997). The Presentation of the Chloride Threshold Level for Corrosion of Steel in Concrete. *Corrosion Science*, 39, 1001–1013.

Glass, G.K., and Buenfeld, N.R. (2000). The Inhibitive Effects of Electrochemical Treatment Applied to Steel in Concrete. *Corrosion Science*, 42, 923–927.

Liu, G., and Gjørv, O.E. (2004). Early Age Resistance of Concrete against Chloride Penetration. In *Proceedings, Fourth International Conference on Concrete under Severe Conditions—Environment and Loading*, vol. 1, ed. B.H. Oh, K. Sakai, O.E. Gjørv, and N. Banthia. Seoul National University and Korea Concrete Institute, Seoul, pp. 165–172.

Malhotra, V.M., Ramachandran, V.S., Feldman, R.F., and Aïtcin, P.C. (1987). *Condensed Silica Fume in Concrete*. CRC Press, Boca Raton, FL.

Malhotra, V.M., and Ramezanianpour, A.A. (1994). *Fly Ash in Concrete*. CANMET, Ottawa.

Mehta, P.K. (1977). *Effect of Cement Composition on Corrosion of Reinforcing Steel in Concrete*, ASTM STP 629, p. 12.

NORDTEST. (1989). *NT Build 355: Concrete, Repairing Materials and Protective Coating: Diffusion Cell Method, Chloride Permeability*. NORDTEST, Espoo, Finland.

NORDTEST. (1995). *NT Build 443: Concrete, Hardened: Accelerated Chloride Penetration*. NORDTEST, Espoo, Finland.

NORDTEST. (1999). *NT Build 492: Concrete, Mortar and Cement Based Repair Materials, Chloride Migration Coefficient from Non-Steady State Migration Experiments*. NORDTEST, Espoo, Finland.

Østvik, J.M. (2005). Thermal Aspects of Corrosion of Steel in Concrete, Ph.D. Thesis 2005:5. Department of Structural Engineering, Norwegian University of Science and Technology—NTNU, Trondheim.

Page, C.L., and Vennesland, Ø. (1983). Pore Solution Composition and Chloride Binding Capacity of Silica-Fume Cement Pastes. *Materials and Structures, RILEM*, 16(91), 19–25.

Poulsen, E., and Mejlbro, L. (2006). *Diffusion of Chlorides in Concrete—Theory and Application*. Taylor & Francis, London.

Sengul, Ö. (2005). Effects of Pozzolanic Materials on Mechanical Properties and Chloride Diffusivity of Concrete, Ph.D. Thesis. Istanbul Technical University, Institute of Science and Technology, Istanbul.

Sengul, Ö., and Gjørv, O.E. (2007). Effect of Blast Furnace Slag for Increased Concrete Sustainability. In *Proceedings, International Symposium on Sustainability in the Cement and Concrete Industry*, ed. S. Jacobsen, P. Jahren, and K.O. Kjellsen. Norwegian Concrete Association, Oslo, pp. 248–256.

Shalon, R., and Raphael, M. (1959). Influence of Seawater on Corrosion of Reinforcement. *Proceedings, ACI*, 55, 1251–1268.

Sluijter, W.L. (1973). *De binding van chloride door cement en de indringsnelheid van chloride in mortel*, IBBC-TNO Report BI-73-41/01.1.310. Delft (in Dutch).

Tang, L., Nilsson, L.-O., and Basher, P.A.M. (2012). *Resistance of Concrete to Chloride Ingress—Testing and Modelling*. Spon Press, London.

Teng, S., and Gjørv, O.E. (2013). Concrete Infrastructures for the Underwater City of the Future. In *Proceedings, Seventh International Conference on Concrete under Severe Conditions—Environment and Loading*, ed. Z.J. Li, W. Sun, C.W. Miao, K. Sakai, O.E. Gjørv, and N. Banthia. RILEM, Bagneux, pp. 1372–1385.

Theissing, E.M., Wardenier, P., and deWind, G. (1975). *The Combination of Sodium Chloride and Calcium Chloride by Some Hardened Cement Pastes*, Stevin Laboratory Report. Delft University of Technology, Delft.

Thomas, M.D.A., Bremner, T., and Scott, A.C.N. (2011). Actual and Modeled Performance in a Tidal Zone. *Concrete International*, 33(11), 23–28.

Trætteberg, A. (1977). *The Mechanism of Chloride Penetration into Concrete*, SINTEF Report STF65 A77070. Trondheim (in Norwegian).

Tuutti, K. (1982). *Corrosion of Steel in Concrete*, Report 4-82. Cement and Concrete Research Institute, Stockholm.

Vennesland, Ø., and Gjørv, O.E. (1981). Effect of Cracks on Steel Corrosion in Submerged Concrete Sea Structures. *Materials Performance*, 20, 49–51.

Vennesland, Ø., and Gjørv, O.E. (1983). Silica Concrete—Protection against Corrosion of Embedded Steel. In *Proceedings, CANMET/ACI International Conference on Fly Ash, Silica Fume and Natural Pozzolans in Concrete*, vol. II, ACI SP-79, ed. V.M. Malhotra, pp. 719–729.

Zhang, T., and Gjørv, O.E. (1996). Diffusion Behavior of Chloride Ions in Concrete. *Cement and Concrete Research*, 26(6), 907–917.

Chapter 4

Durability analysis

4.1 GENERAL

Depending on the quality of the concrete and the thickness of the concrete cover, it may take many years before the chlorides reach embedded steel and corrosion starts. After the chlorides have reached the embedded steel, however, it may only take a few years before visual damage in the form of cracks and rust staining appears, but it may still take a long time before the load carrying capacity of the structure is severely reduced, as outlined and discussed in Chapter 2. Schematically, the deteriorating process takes place as shown in Figure 4.1. As soon as corrosion starts, a very complex system of galvanic cell activities in the concrete structure develops. In this system of galvanic cell activities, the deterioration appears in the form of concentrated pitting corrosion in the anodic areas of the rebar system, while the adjacent cathodic areas act as catchment areas for oxygen. Although larger portions of the rebar system eventually become depassivated, not all of these areas will necessarily corrode. As already discussed in Chapter 3, the steel in the first and most active corroding parts of the structure may act as sacrificial anodes, and thus cathodically protect the other parts of the structure. Since structural shape, electrical continuity, and local exposure conditions will affect this pattern of deterioration, it appears very difficult to develop a general mathematical model for predicting the time necessary before the load carrying capacity of the structure becomes reduced. Although several attempts for developing such a mathematical model have been made (Lu et al., 2008), it appears that no reliable mathematical model or numerical solution for this very complex deteriorating process currently exists. Already in the early 1970s, however, Collepardi et al. (1970, 1972) came up with a relatively simple mathematical model for estimating the time necessary for the chlorides to reach embedded steel through the concrete cover of a given quality and thickness in a given environment.

Although it is possible to roughly estimate the time necessary before the chlorides reach embedded steel and corrosion starts, this does not provide any basis for prediction or estimation of the service life of the structure. As

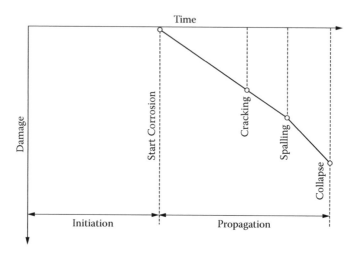

Figure 4.1 Deterioration of a concrete structure due to steel corrosion. (From Tuutti, K., *Corrosion of Steel in Concrete*, Report 4, Swedish Cement and Concrete Institute, Stockholm, Sweden, 1982.)

soon as corrosion starts, however, the owner of the structure has gotten a problem. At an early stage of visual damage, this only represents a maintenance and cost problem, but later on, it may gradually develop into a more difficult controllable safety problem. As a basis for the durability design, therefore, efforts should be made to obtain a best possible control of the chloride ingress during the initiation period before any corrosion starts. It is in this early stage of the deteriorating process that it is both technically easier and much cheaper to take necessary precautions and select proper protective measures for control of the further deteriorating process. Control of the chloride ingress during the initiation period in the form of a proper durability design and preventive maintenance has also shown to be a very good strategy from a sustainability point of view (Chapter 11).

Since all the input parameters necessary for calculating the rate of chloride ingress through the concrete cover always show a high scatter and variability, it is very appropriate to combine this calculation with a probability analysis that can take some of this scatter and variability into account (DuraCrete, 2000). In this way, it is possible to estimate the probability for a critical amount of chlorides to reach the embedded steel during a certain service period for the given concrete structure in the given environment.

For such a probability-based durability design, a serviceability limit state (SLS) must also be defined. Although different stages of the deteriorating process may be chosen as a basis for such a serviceability limit state, the onset of steel corrosion is a very critical stage that has been chosen as a proper serviceability limit state in the following.

In recent years, a rapid international development of models and procedures for probability-based durability design of concrete structures has taken place (Siemes and Rostam, 1996; Engelund and Sørensen, 1998; Gehlen, 2000; DuraCrete, 2000; FIB, 2006, Tang et al., 2012), and in many countries, such durability design has been applied to a number of important concrete structures (Stewart and Rosowsky, 1998; McGee, 1999; Gehlen and Schiessl, 1999; Gehlen, 2007). Also, in Norway such durability design has been applied to a number of concrete structures where high safety, durability, and service life have been of special importance (Gjørv, 2002, 2004). In the beginning, this design was primarily based on the results and guidelines from the European research project DuraCrete (2000), but successively, as practical experience with such design was gained, the basis for the design was simplified and further developed for more practical applications, but the basic principles remained essentially the same. Thus in 2004, this design was adopted by the Norwegian Association for Harbor Engineers as the basis for new recommendations and guidelines for construction of more durable concrete infrastructure in Norwegian harbors (NAHE, 2004a, 2004b, 2004c). Lessons learned from practical experience with these recommendations were incorporated into subsequent revised editions, the third and last of which from 2009 was also adopted by the Norwegian Chapter of PIANC, which is the World Association for Waterborne Transport Infrastructure (PIANC/NAHE, 2009a, 2009b, 2009c).

In the following, the basis for calculation of both chloride ingress and probability of corrosion is briefly described, and then the necessary input parameters for the durability analysis are outlined and discussed. In order to demonstrate how calculations of corrosion probability can be applied as a basis for the durability design, two case studies are also briefly outlined and discussed.

4.2 CALCULATION OF CHLORIDE INGRESS

As already discussed in Chapter 3, rather complex transport mechanisms for the ingress of chlorides into concrete exist. In a very simplified form, however, the rate of chloride ingress can be estimated by use of Fick's second law of diffusion according to Collepardi et al. (1970, 1972), in combination with a time-dependent chloride diffusion coefficient according to Takewaka and Mastumoto (1988) and Tang and Gulikers (2007), as shown in Equations 4.1 and 4.2:

$$C(x,t) = C_S \left[1 - erf\left(\frac{x_C}{2\sqrt{D(t) \cdot .t}} \right) \right] \tag{4.1}$$

In this equation, $C(x,t)$ is the chloride concentration in depth x_C after time t, C_S is the chloride concentration at the concrete surface, D is the concrete chloride diffusion coefficient, and erf is a mathematical function.

$$D(t) = \frac{D_0}{1-\alpha} \left[\left(1+\frac{t'}{t}\right)^{1-\alpha} - \left(\frac{t'}{t}\right)^{1-\alpha} \right] \left(\frac{t_0}{t}\right)^\alpha \cdot k_e \tag{4.2}$$

In this equation, D_0 is the diffusion coefficient after the reference time t_0, and t' is the age of concrete at the time of chloride exposure. The parameter α represents the time dependence of the diffusion coefficient, while k_e is a parameter that takes the effect of temperature into account according to Kong et al. (2002):

$$k_e = exp\left[\frac{E_A}{R}\left(\frac{1}{293} - \frac{1}{273+T} \right) \right] \tag{4.3}$$

where exp is the exponential function, E_A is the activation energy for the chloride diffusion, R is the Gas constant, and T is the temperature.

The criterion for steel corrosion then becomes

$$C(x) = C_{CR} \tag{4.4}$$

where $C(x)$ is the chloride concentration at the depth of the embedded steel, and C_{CR} is the critical chloride concentration in the concrete necessary for depassivation and onset of corrosion.

4.3 CALCULATION OF PROBABILITY

For the structural design of concrete structures, the main objective is always to establish the combined effects of external loads (S) and the resistance to withstand these loads (R) in such a way that the design criterion becomes

$$R \geq S \text{ or } R - S \geq 0 \tag{4.5}$$

When $R < S$, failure will occur, but since all factors affecting R and S always show a high scatter and variability, all established procedures for structural design have properly taken this into account.

In principle, the durability design takes the same approach as that of the structural design. In this case, however, the loads (S) are the combined effects of both chloride loads and temperature conditions, while the resistance to withstand these loads (R), which is the resistance to chloride

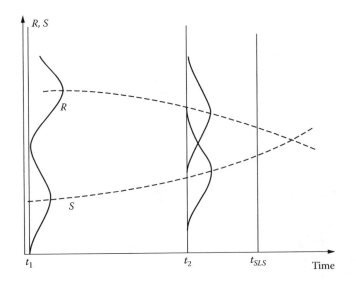

Figure 4.2 The principles of a time-dependent reliability analysis.

ingress, is the combined effects of both concrete quality and concrete cover. Although neither S nor R is comparable to that of structural design, the acceptance criterion for having a probability of failure less than a given value is the same.

In Figure 4.2, the scatter and variability of both R and S are demonstrated in the form of the two distribution curves along the y-axis. At an early stage, there is no overlapping between these two distribution curves, but over time, a gradual overlapping from time t_1 to t_2 takes place. This increasing overlapping will at any time reflect the probability of failure or the probability for onset of steel corrosion, and gradually, the upper acceptable level for the probability of failure (t_{SLS}) is reached and exceeded.

In principle, the probability of failure can be written as

$$p(failure) = p_f = p(R - S < 0) < p_0 \tag{4.6}$$

where p_0 is a measure for failure probability.

For the structural design, the safety of the structure is normally expressed in the form of both reliability and a reliability index for assessment of the possible consequences of the failure for the structural safety. Since there are no immediate consequences of an onset of corrosion, however, there is no reliability index to be considered in the further procedure. In the current codes for reliability of structures, however, an upper level for probability of failure of 10% in the serviceability limit state is often specified (Standard Norway, 2004). Also for the onset of corrosion, therefore, an upper probability level of 10% has been adopted as a basis for the further durability design. Normally, the above failure function includes a number of variables, all of which have their own statistical parameters. Therefore, the use of such a failure function requires numerical calculations and the application of special software. Currently, there are several mathematical methods available for evaluation of the failure function, such as

- First-order reliability method (FORM)
- Second-order reliability method (SORM)
- Monte Carlo simulation (MCS)

4.4 CALCULATION OF CORROSION PROBABILITY

In principle, the calculation of corrosion probability can be carried out by use of any of the above mathematical methods. In the following, however, the calculation of corrosion probability is based on the modified Fick's second law of diffusion in Equation 4.1, in combination with a Monte Carlo simulation. Although such a combined calculation can also be carried out in different ways, a special software DURACON for this calculation was developed (Ferreira, 2004; Ferreira et al., 2004, http://www.pianc.no/duracon.php).

A Monte Carlo simulation can briefly be described as a statistical simulation method where sequences of random numbers are applied to perform the simulation. When the ingress of chlorides is simulated by use of Equation 4.1, this requires that all the input parameters to this equation can be described by a probability density function. Once these functions of the various durability parameters of the system are known, the probability of failure is based on the evaluation of the limit state function for a large number of simulations. The failure function is then calculated for each outcome. If the outcome is in the failure region, then a contribution to the probability of failure is obtained.

For the DURACON software, the concrete cover was chosen as the resistance variable (r), while the depth of the critical chloride front was chosen as the load variable (s). The probability of failure is then estimated by use of the following expression:

$$p_f = \frac{1}{N} \cdot \sum_{j=1}^{N} I\left[g\left(r_j, s_j\right)\right] \tag{4.7}$$

where N is the number of simulations, $I[g(r_j, s_j)]$ is the indicator function, and $g(r_j, s_j)$ is the limit state equation; s represents the environmental load, and r represents the resistance of the concrete to chloride ingress.

The standard error of the above calculation can be estimated by the following expression according to Thomas and Bamforth (1999):

$$s = \sqrt{\frac{p_f\left(1-p_f\right)}{N}} \tag{4.8}$$

from which it can be seen that the accuracy of the Monte Carlo simulation mainly depends on the number of simulations.

Based on the above calculation of corrosion probability, a certain service period can be specified before the corrosion probability of 10% is reached. This provides the basis for the durability design of new concrete structures. In this design, a comparison and selection of one of several technical solutions for the given concrete structure in the given environment for the specified service period are made.

It should be noted, however, that the above calculation is based on a number of assumptions and simplifications. Therefore, for increased service periods of more than 100 years, the calculation of corrosion probability gradually becomes less reliable. For practical applications, the corrosion probability should be kept as low as possible and not exceed 10% for service periods of up to 150 years, but in addition, some further protective measures, such as partial use of stainless steel, should also preferably be applied. For service periods of more than 150 years, however, any calculations of corrosion probability are no longer considered valid. For such long service periods, the corrosion probability should still be kept as low as possible and not exceed 10% for a 150-year service period, but in addition, one or more additional protective measures should always be applied in order to further increase and ensure the durability.

Based on the calculations of corrosion probability for the specified service period, requirements to both concrete quality (chloride diffusivity) and concrete cover are established. During concrete construction, this provides the basis for the concrete quality control and quality assurance (Chapter 6). Upon completion of the concrete construction, a new calculation of corrosion probability based on the achieved data on chloride diffusivity and concrete cover is carried out. This provides the basis for documenting the achieved construction quality and compliance with the specified durability (Chapter 7). During operation of the structure, calculations of corrosion probability are further carried out as a basis for the condition assessment and preventive maintenance (Chapter 8). In this case, the calculations of corrosion probability are based on the observed rates of chloride ingress in the form of the apparent chloride diffusivities in combination with the previously observed data on achieved concrete cover.

For all the above types of probability calculation, certain input parameters to the calculations are needed. For the durability design, all the necessary input parameters are described and discussed in the following. For the documentation of achieved construction quality and compliance with the specified durability, the input parameters are described and discussed in Chapters 6 and 7, while for the condition assessment and preventive maintenance during operation of the structure, the input parameters are described and discussed in Chapter 8.

4.5 INPUT PARAMETERS

4.5.1 General

In general, the durability design should always be an integral part of the structural design of the given structure. Already at an early stage of the design, therefore, a certain service period should be required before 10% probability of corrosion is reached. Before a proper technical solution is selected, it may be necessary to carry out several calculations for various combinations of possible concrete qualities and concrete covers.

For all calculations of corrosion probability, proper information about the following input parameters is needed:

- Environmental loading
  - *Chloride loading, C_S*
  - *Age at chloride loading, t'*
  - *Temperature, T*

- Concrete quality
  - *Chloride diffusivity, D*
  - *Time dependence factor, α*
  - *Critical chloride content, C_{CR}*

- Concrete cover, X_C

It should be noted that the above input parameters to the durability design may have different distribution characteristics. If nothing else is known for the distribution of the various input parameters, however, a normal distribution with a coefficient of variation between 0.1 and 0.2 may be assumed. For the documentation of achieved construction quality, the input parameters on both chloride diffusivity and concrete cover are based on the obtained data from the quality control during concrete construction. Since these data may occasionally show a high scatter and variability, other probability distributions for these input parameters, such as beta-distribution, may be assumed. The same is true for the durability analyses carried out as a basis for the condition assessment during operation of the structures. In this case, the input parameter on chloride diffusivity is based on the observed data from the real chloride ingress taking place, which may also show a high scatter and variability.

In the following, however, some general guidelines for determination and selection of the above input parameters used for the durability design are given.

4.5.2 Environmental loading

4.5.2.1 Chloride loading, C_S

For all concrete structures in chloride-containing environments, the chloride loading is normally defined as the accumulated surface chloride concentration on the concrete surface (C_S) after some time of exposure (Figure 4.3). This chloride ingress curve is the result of a regression analysis of at least six observed data on the chloride ingress after a given time of exposure and curve fitting to Fick's second law. The surface chloride concentration (C_S), which is normally higher than the maximum observed chloride concentration in the surface layer of the concrete (C_{max}), is primarily the result of the local environmental exposure, but concrete quality, geometrical shape of the structure, and height above water also affect the accumulation of

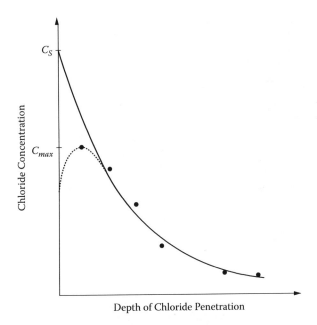

Figure 4.3 Definition of the surface chloride concentration (C_S) based on a regression analysis of observed data on chloride ingress.

the surface chloride concentration. For all concrete structures, therefore, the accumulated surface chloride concentration typically shows a very high scatter and variability, as shown in Chapter 2. For the durability analysis, however, it is important to estimate and select a proper value of the surface chloride concentration (C_S) that is as representative as possible for the most exposed and critical parts of the structure. In some cases, it may be appropriate to select different chloride loads for different parts of the structure and then carry out separate probability calculations for the various parts of the structure.

For a new concrete structure, it may not be easy to estimate and select a proper value for the chloride loading as generally described above. If possible, therefore, data from previous field investigations of similar types of concrete structures in similar types of environments should be applied. In many countries, a large number of both concrete bridges and concrete harbor structures in severe marine environments have been the subject of extensive field investigations. Thus, along the Norwegian coastline, extensive measurements of chloride ingress have been carried out on a large number of concrete structures, as previously described in Chapter 2, some data from which have been plotted in Figures 4.4 and 4.5. For the individual

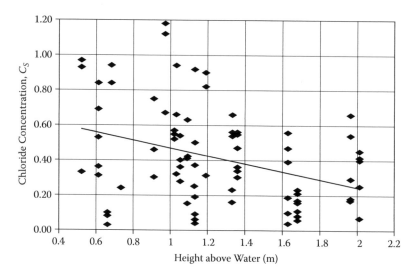

Figure 4.4 Obtained surface chloride concentrations by weight of concrete (C_S) on Norwegian concrete harbor structures. (From Markeset, G., Service Life Design of Concrete Structures Viewed from an Owner's Point of View, in *Proceedings, Seminar on Service Life Design of Concrete Structures*, Norwegian Concrete Association, Oslo, Norway, 2004, pp. 13.1–13.30 [in Norwegian]; based on data from Hofsøy, A. et al., *Experiences from Concrete Harbour Structures*, Report 2.2, Research Project Durable Concrete Structures, Norwegian Public Road Administration, Oslo, Norway, 1999 [in Norwegian].)

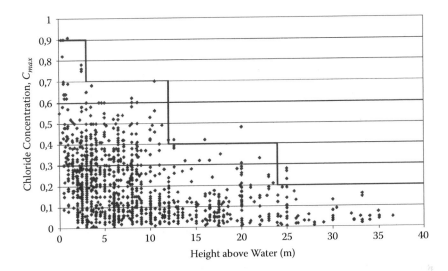

Figure 4.5 Observed maximum surface chloride concentrations by weight of concrete (C_{max}) on Norwegian concrete coastal bridges. (From Fluge, F., Marine Chlorides—A Probabilistic Approach to Derive Provisions for EN 206-1, in *Proceedings, Third Workshop on Service Life Design of Concrete Structures—From Theory to Standardisation*, DuraNet, Tromsø, Norway, 2001, pp. 47–68.)

concrete structures, experience has shown that the surface chloride concentrations (C_S) successively accumulate over a certain number of years before they tend to level out to fairly stable values for the given environmental exposure. In the following, therefore, the chloride loading (C_S) is assumed to be a constant input parameter for the given environment.

Although the selection of chloride loads for a new concrete structure should preferably be based on local experience from similar concrete structures exposed to similar environments, general experience available from the literature may also provide a basis for the assessment of a proper chloride loading. Based on general experience from concrete structures in marine environments, some general guidelines for estimation of chloride loading are given in Table 4.1. Since the data on chloride concentrations

Table 4.1 General guidelines for estimation of chloride loading (C_S) on concrete structures in marine environments

Environmental load	C_S (% by wt. of cement)	
	Mean value	*Standard deviation*
High	5.5	1.3
Average	3.5	0.8
Moderate	1.5	0.5

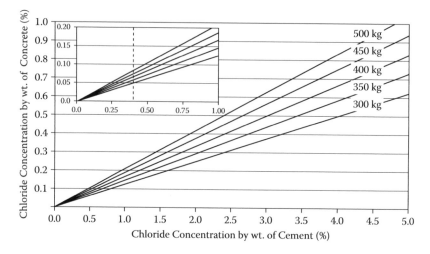

Figure 4.6 Conversion diagram for estimating chloride concentrations in % by weight of cement based on % by weight of concrete with various cement contents. (From Ferreira, M., Probability Based Durability Design of Concrete Structures in Marine Environment, Doctoral Dissertation, Department of Civil Engineering, University of Minho, Guimarães, Portugal, 2004.)

are often given in % by weight of concrete, a general conversion diagram to chloride concentrations in % by weight of cement is also given (Figure 4.6).

4.5.2.2 Age at chloride loading, t′

Since the resistance of the given concrete to chloride ingress very much depends on the degree of hydration, the age of the concrete at the time of chloride loading (t') is also a very important parameter for the assessment of chloride ingress. A proper selection of this input parameter further depends on both type of concrete and construction procedure, local curing conditions, and risk for early-age chloride exposure. Occasionally, a very early-age chloride exposure during concrete construction may take place, as shown in Chapter 2.

4.5.2.3 Temperature, T

For a given concrete structure in a given environment, the rate of chloride ingress also very much depends on the temperature, as shown in Equation 4.3. Based on local information on prevailing temperature conditions, data on average annual temperatures may be used as a basis for the selection of this input parameter.

4.5.3 Concrete quality

4.5.3.1 Chloride diffusivity, D

As discussed in Chapter 3, the chloride diffusivity (D) of a given concrete is a very important quality parameter that generally reflects the resistance of the concrete to chloride ingress. Although the water/binder ratio also reflects the porosity, and hence the resistance to chloride ingress, extensive experience demonstrates that the selection of a proper binder system may be more important for obtaining a high resistance to chloride ingress than selecting a low water/binder ratio. Thus, when the water/binder ratio was reduced from 0.50 to 0.40 for a concrete based on a pure portland cement, the chloride diffusivity was only reduced by a factor of two or three, while incorporation of various types of supplementary cementitious materials, such as blast furnace slag, fly ash, or silica fume, at the same water/binder ratio reduced the chloride diffusivity by a factor of up to 20 (Thomas et al., 2011). While a reduced water/binder ratio from 0.45 to 0.35 for a concrete based on a pure portland cement may only reduce the chloride diffusivity by a factor of two, a replacement of the portland cement by a proper blast furnace slag cement may reduce the chloride diffusivity by a factor of up to 50 (Bijen, 1998). By also combining the blast furnace slag cement with silica fume, very low chloride diffusivities can be obtained, and hence a concrete with a very high resistance against chloride ingress can be produced (Chapter 3).

In the literature, there are several types and definitions of the chloride diffusivity of a given concrete as well as several methods for testing of the chloride diffusivity (Schiessl and Lay, 2005; Tang et al., 2012). Thus, NORDTEST has standardized three different types of test method, including the steady-state migration method NT Build 355 (NORDTEST, 1989) the immersion test method NT Build 443 (NORDTEST, 1995), and the non-steady-state migration method NT Build 492 (NORDTEST, 1999). All of these test methods are accelerated methods giving different values for the chloride diffusivity, but since they all show a good correlation, any of the above test methods can be used for both quantifying and comparing the resistance to chloride ingress of various types of concrete (Tong and Gjørv, 2001; Schiessl and Lay, 2005; Tang et al., 2012).

Although all the above types of test method are accelerated test methods, the duration of the testing is very different for the different test methods. Both the steady-state migration method and the immersion test method are based on well-cured concrete specimens before exposure, and the testing may take a long time. Since the non-steady-state migration method does not require any precuring, however, this is the only method for a very rapid testing, independent of concrete age. Therefore, in order to be applicable for a regular quality control during concrete construction, the non-steady-state migration method or the so-called rapid chloride migration (RCM) method was adopted; in particular, this has shown to be a very appropriate method when it is also combined with a corresponding testing of the electrical resistivity of the given concrete (Gjørv, 2003), as further described and discussed for concrete quality control in Chapter 6.

The RCM method, which was originally developed by Tang in 1996 (Tang, 1996a, 1996b), was later the subject for extensive testing and comparison with other test methods in the European research project DuraCrete (2000). As a result, the chloride diffusivity based on the RCM method was adopted as a basis for the general guidelines for durability design developed in this research project. A very good correlation with the results obtained by the RCM method and the steady-state migration method was also reported by Tong and Gjørv (2001). The strong statistical correlation with the results obtained by the immersion test method can also be seen in Figure 4.7.

Already in 2001, a very good documentation on the precision of the RCM method was published by Tang and Sørensen (2001). Several test methods were also compared and evaluated in the European research

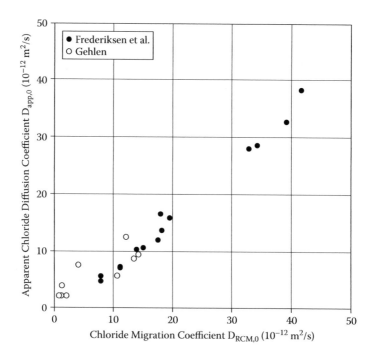

Figure 4.7 Correlation of the effective diffusion coefficients obtained by immersion tests and the RCM method according to Gehlen (2000) and Fredriksen et al. (1996). (From Schiessl, P., and Lay, S., Influence of Concrete Composition, in Corrosion in Reinforced Concrete Structures, ed. H. Böhni, Woodhead Publishing, Cambridge, England, 2005, pp. 91–134.)

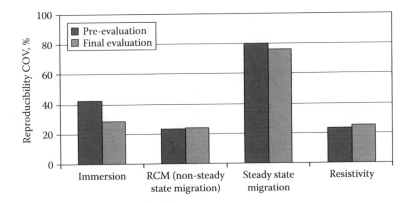

Figure 4.8 Precision among all the test methods for chloride diffusivity evaluated in the European research project ChlorTest (2005).

project ChlorTest (2005), in which it was shown that the RCM method gave the best precision (Figure 4.8). Due to its simplicity, rapidity, and precision, the RCM method has gradually also been adopted internationally (Hooton et al., 2000, AASHTO, 2003); in the United States, it has been adopted as AASHTO TP 64-03, while in China it has been adopted as the Chinese national standard GB/T 50082. In the following, the chloride diffusivity (D) of the concrete is both defined and based on the RCM method NT Build 492 (NORDTEST, 1999).

For most durability analyses, the 28-day RCM diffusivity is normally used as an input parameter to the durability design in the same way as the 28-day compressive strength is used as an input parameter to the structural design. However, additional durability analyses based on chloride diffusivities obtained after longer curing periods can also be carried out, and this may be appropriate if some of the types of concrete are based on binder systems that hydrate very slowly, such as blended cements based on fly ash. However, for the regular concrete quality control during concrete construction and the documentation of compliance with the specified durability, the testing is normally based on the 28-day chloride diffusivity (D_{28}), as further described and discussed under concrete quality control in Chapter 6.

For continued water curing at 20°C beyond 28 days in the laboratory, the chloride diffusivity is successively reduced over a certain period of time, but somewhat depending on type of binder system, it typically tends to level out during a curing period of approximately one year. Therefore, the obtained chloride diffusivity after one year of water curing at 20°C (D_{365}) is used as a basis for characterizing the potential resistance of the given concrete to chloride ingress, as described and discussed under documentation of achieved construction quality in Chapter 7.

Table 4.2 Resistance to chloride ingress of various types of concrete based on the 28-day RCM diffusivity

Chloride diffusivity, D_{28} m²/s × 10^{-12}	Resistance to chloride ingress
>15	Low
10–15	Moderate
5–10	High
2.5–5	Very high
<2.5	Extremely high

Source: Nilsson, L. et al., High-Performance Repair Materials for Concrete Structures in the Port of Gothenburg, in Proceedings, Second International Conference on Concrete under Severe Conditions— Environment and Loading, vol. 2, ed. O.E. Gjørv et al., E & FN Spon, London, England, 1998, pp. 1193–1198.

As a basis for a more general assessment of the resistance to chloride ingress of various types of concrete based on the 28-day RCM diffusivity, some general values are shown in Table 4.2.

4.5.3.2 Time dependence factor, α

Since the chloride diffusivity is a time-dependent property of the concrete, this time dependence (α) is also a very important input parameter generally reflecting how the chloride diffusivity of a given concrete in a given environment develops over time. In order to select a proper α-value, an empirical value for the given type of concrete in the given type of environment is normally used as an input parameter to the durability analyses.

For new concrete structures, therefore, the same problem to select a proper α-value exists as that already discussed for the selection of a proper chloride loading (C_S). Again, current experience from field investigations of similar concrete structures in similar environments may provide a basis for selecting a proper α-value. Also, information from long-term field tests with similar types of concrete in similar environments may be available from the literature. Based on such experience, some general guidelines for selecting a proper α-value are given in Table 4.3. This table shows some α-values obtained for concrete based on various types of cement exposed to the tidal and splash zone of marine environments (Mangat and Molloy, 1994; Thomas and Bamforth, 1999; Thomas et al., 1999; Bamforth, 1999). Although combinations of the various types of cement with supplementary cementitious materials such as silica fume always will reduce the chloride diffusivity, current experience indicates that Table 4.3 may still be used as a general and rough basis for estimation of a proper α-value.

Table 4.3 General guidelines for estimation of α-values for tidal and splash zone exposure of concrete structures in marine environments

Concrete based on various types of cement	α-value	
	Mean value	Standard Deviation
Portland cements	0.4	0.08
Blast furnace slag cements	0.5	0.10
Fly ash cements	0.6	0.12

4.5.3.3 Critical chloride content, C_{CR}

As already discussed in Chapter 3, a number of factors affect the depassivation of embedded steel in concrete; depending on all these factors, the critical chloride concentration in the pore solution may vary within wide limits. Also, due to the very complex relationship between the total chloride content in the concrete and the critical chloride concentration in the pore solution, it is not possible to give any general values for the critical chloride content. When certain values for the critical chloride content nevertheless are given in existing concrete codes and recommendations, this is only based on empirical information on total chloride contents in the concrete, which may give a certain risk for development of corrosion. For traditional carbon steel, some empirical values are shown in Table 4.4. However, whether steel corrosion will develop or not also depends on other corrosion parameters, such as oxygen availability and electrical resistivity of the concrete. Thus, the risk for development of corrosion may be very low both in wet or submerged concrete due to very low oxygen availability and in very dry concrete due to ohmic control of the corrosion process (Figure 4.9).

Based on empirical information from a wide range of concrete qualities and moisture conditions, an average value for critical chloride content

Table 4.4 Risk for development of corrosion of carbon steel depending on total chloride content

Chloride content (%)		Risk of corrosion
By wt. of cement	By wt. of concrete[a]	
>2.0	>0.36	Certain
1.0–2.0	0.18–0.36	Probable
0.4–1.0	0.07–0.18	Possible
<0.4	<0.07	Negligible

Source: Browne, R. et al., Marine Durability Survey of the Tongue Sand Tower, Concrete in the Ocean Program, CIRIA UEG Technical Report 5, Cement and Concrete Association, London, England, 1980.

[a] Based on 440 kg/m³ of cement.

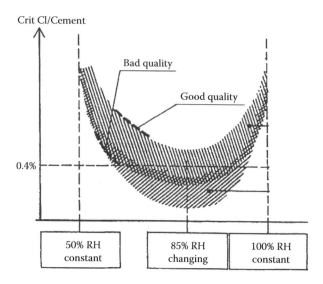

Figure 4.9 Qualitative relationship between critical chloride content (C_{CR}), environmental conditions, and quality of concrete. (From CEB, *Durable Concrete Structures—Design Guide*, Comité Euro-International du Beton—CEB, Bulletin D'Information 183, Thomas Telford, London, England, 1992.)

of 0.4% by weight of cement for reinforcement based on carbon steel is often specified in current concrete codes. If nothing else is known, therefore, an average value of 0.4% with a standard deviation of 0.1% by weight of cement may be selected as an input parameter to the durability analysis. It should be noted, however, that the above level of 0.4% is typically based on previous experience with moderate concrete qualities and pure portland cements; higher concrete qualities with typically higher levels of electrical resistivities may show somewhat higher values of the critical chloride content.

For more corrosion-sensitive types of high-quality steel, an average value of 0.1% with a standard deviation of 0.03% may be selected. For various grades of stainless steel reinforcement, the critical chloride content may typically vary from 2.5 to 3.5% by weight of cement, but grades with threshold values of up to 5 to 8% are also available (Chapter 5).

4.5.4 Concrete cover, X_C

In current concrete codes, requirements to both minimum concrete cover ($X_{C,min}$) and tolerances are given for the given environment. Thus, the nominal concrete cover ($X_{C,N}$) is always specified with a certain value of tolerance (ΔX_C), and different values for ΔX_C may be specified. For a tolerance of ±10 mm, the minimum requirement to concrete cover then becomes

$$X_{C,min} = X_{C,N} - 10 \qquad (4.9)$$

Although the specified concrete cover primarily gives the required concrete cover to the structural steel, additional mounting steel for ensuring the position of the structural steel during concrete construction is also often applied. Since the penetrating chlorides do not distinguish between structural and mounting steel, however, the nominal concrete cover should preferably be specified for all embedded steel, including the mounting steel, in order to avoid any cracking of the concrete cover due to premature corrosion. As part of the structural design, great efforts to avoid any cracking of the concrete are generally made. Cracking of the concrete cover caused by corroding mounting steel may represent the same type of weakness as that caused by any other types of cracking. Therefore, instead of using mounting steel, which can corrode, mounting systems based on noncorroding materials such as those discussed in Chapter 5 should preferably be applied.

If it is assumed that 5% of the reinforcing steel has a concrete cover less than $X_{C,min}$, the durability analysis can be based on an average concrete cover of $X_{C,N}$ with a standard deviation of $\Delta X_C/1.645$. Then, the effect of increased concrete cover beyond that required in current concrete codes can be quantified. For the documentation of achieved construction quality, as described later in Chapter 5, however, the durability analyses must always be based on the obtained values for both concrete cover and standard deviation from the regular quality control during concrete construction.

4.6 CASE STUDIES

4.6.1 General

In order to demonstrate how the above calculations of corrosion probability can be applied as a basis for the durability design of new concrete structures, two different types of case study are briefly outlined and discussed in the following.

For a new concrete harbor structure in a marine environment with a typical annual temperature of 10°C, the overall durability requirement was based on a 120-year service period before the 10% probability level of corrosion would be reached. In this case, durability analyses were carried out in order to select a proper combination of concrete quality and concrete cover that would meet this durability requirement. In order to demonstrate the importance of the temperature in the environment, some additional durability analyses for increased annual temperatures of 20 and 30°C, respectively, were also carried out.

The other case study is part of the more comprehensive NRF Research Program *Underwater Infrastructure and Underwater City of the Future* at

Nanyang Technological University in Singapore (NTU, 2011). The objective of this five-year NRF research program, which started in 2011, is to develop the technical basis necessary for a future development of Singapore City based on a large number of sea-spaced concrete structures. For such important concrete infrastructures, a highest possible durability and reliability would be of vital importance, and therefore, a service life as high as possible would be required. Since any calculations of corrosion probability for a service period of more than 150 years are not considered valid, however, a service period of up to 150 years with a corrosion probability as low as possible was selected as a basis for the durability analyses. In order to further increase and ensure the durability, one or more additional protective measures must also be applied as outlined and discussed in Chapter 5.

In both of the above cases, a number of concrete mixtures based on various types of binder system for testing the chloride diffusivity were produced. Based on the results obtained, durability analyses were carried out in order to find out how the different types of concrete would affect the probability of corrosion during the required service periods. In a next step, further durability analyses were carried out in order to find out how different concrete covers also would affect the corrosion probability.

4.6.2 Concrete harbor structure

4.6.2.1 Effect of concrete quality

In order to select a proper concrete quality for the concrete harbor structure, four concrete mixtures (Types 1–4) were produced for which the 28-day chloride diffusivities (D_{28}) were determined. These concrete mixtures were produced with four different types of commercial cement in combination with silica fume. The different types of cement included one high-performance portland cement (CEM I 52.5 LA), one blended portland cement with approximately 20% fly ash (CEM II/A V 42.5 R), and two types of blast furnace slag cements, one with 34% slag (CEM II/B-S 42.5 R NA) and one with 70% slag (CEM III/B 42.5 LH HS), respectively.

Apart from type of binder system, the composition of all the concrete mixtures was the same; they were all produced with 390 kg/m³ of cement in combination with 39 kg/m³ of silica fume (10%), giving a water/binder ratio of 0.38. Thus, all concrete mixtures would fulfill even the strictest durability requirements according to current European Concrete Codes for a 100-year service life (CEN, 2009).

Based on the obtained 28-day chloride diffusivities (D_{28}), an average concrete cover of 70 mm (X_C) with a standard deviation of 6 mm, and estimated values for both the time dependence of the chloride diffusivities (α) and the critical chloride content (C_{CR}), as shown in Table 4.5, durability analyses were carried out. All the other input parameters, including

Table 4.5 Input parameters for analyzing the effect of concrete quality

	Input parameter		
Concrete quality	D_{28} $(m^2/s \times 10^{-12})$	α	C_{CR} (% by wt. of binder)
Type 1 (CEM I 52.5 LA + 10% CSF)	N[a](6.0, 0.64)	N(0.4, 0.08)	
Type 2 (CEM II/A −V 42.5 R + 10% CSF)	N(7.0, 1.09)	N(0.6, 0.12)	
Type 3 (CEM II/B − S 42.5 R NA + 10% CSF)	N(1.9, 0.08)	N(0.5, 0.10	N(0.4, 0.10)
Type 4 (CEM III/B 42.5 LH HS + 10% CSF)	N(1.8, 0.15)		

[a] Normal distribution with average value and standard deviation.

estimated values for both environmental loading C_S (5.5; 1.35%) and age at chloride loading t' (28 days), were kept constant; only the temperature T was increased from 10°C up to 20 and 30°C, respectively.

Although all four concrete mixtures complied with the current European code requirements for a 100-year service life, the service period before the 10% probability of corrosion would be reached differed significantly. At an annual average temperature of 10°C, it can be seen from Figure 4.10 that the concrete based on portland cement in combination with silica fume (Type 1) would give a service period of about 30 years, while the fly ash cement (Type 2) would increase the service period up to about 80 years before the 10% probability of corrosion would be reached. However, only the concrete based on the two types of blast furnace slag cement (Types 3

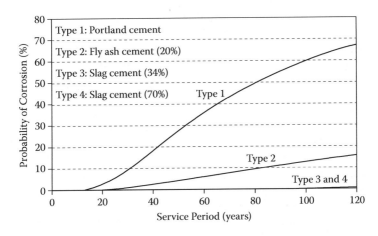

Figure 4.10 Effect of cement type on the probability of corrosion (10°C).

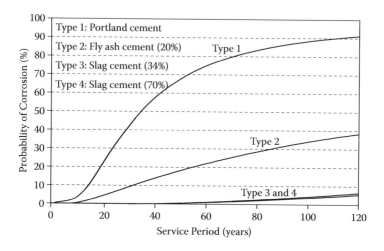

Figure 4.11 Effect of cement type on the probability of corrosion (20°C).

and 4) would meet the specified durability requirement by giving a service period of 120 years or more.

For an increased annual temperature from 10 up to 20°C, it can be seen from Figure 4.11 that the concrete based on portland cement (Type 1) would get a reduced service period from about 30 to less than 20 years, while the concrete based on fly ash cement (Type 2) would get a reduced service period from about 80 to about 30 years. Still, both types of concrete based on blast furnace slag cement (Types 3 and 4) would meet the specified durability requirement by giving a service period of 120 years or more.

By a further increased temperature of up to 30°C, it can be seen from Figure 4.12 that the concrete based on portland cement (Type 1) would only give a service period of less than 10 years, while the concrete based on fly ash cement (Type 2) would give a service period of less than 20 years. At a temperature of 30°C, also the two types of concrete based on slag cement (Types 3 and 4) would no longer meet the required service period of 120 years before 10% probability of corrosion would be reached. These calculations therefore demonstrate that a hot climate with a typical annual temperature of 30°C represents a special challenge to the durability design.

For the above durability analyses, it may be argued that the calculations were only carried out on the basis of the obtained 28-day chloride diffusivities, while the chloride diffusivity of the various types of concrete based on the various types of binder system would develop very differently. Therefore, some additional durability analyses based on values of the chloride diffusivity obtained after longer curing periods were also carried out, but none significantly changed the relative basis for selecting the best concrete from a durability point of view.

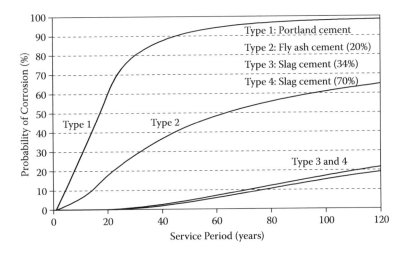

Figure 4.12 Effect of cement type on the probability of corrosion (30°C).

4.6.2.2 Effect of concrete cover

In order to further evaluate the effect of increased concrete cover beyond the nominal cover of 70 mm, a next step of durability analyses based on 90 and 120 mm concrete cover was carried out (Table 4.6). These analyses were only based on the above type concrete with portland cement (Type 1) at a temperature of 10°C, while holding all the other input parameters constant. In order to investigate the effect of increased temperature, however, some additional analyses were also carried out at a temperature of 20°C.

As clearly demonstrated in Figure 4.13, an increased concrete cover would also significantly affect the obtained probability of corrosion. While a nominal cover of 70 mm for a concrete quality of Type 1 would only give a service period of about 30 years, an increased concrete cover of up to 90 and 120 mm would increase the service period of up to about 60 and more

Table 4.6 Input parameters for analyzing the effect of concrete cover

Input parameter	Average value	Standard deviation	Comments
D_{28}	6.0	0.64	Chloride diffusivity (m²/s × 10⁻¹²)
α	0.4	0.08	Time dependence factor
C_{CR}	0.4	0.10	Critical chloride content (% by wt. of binder)
x_C	70	6	Nominal concrete cover (mm)
	90	6	
	120	6	

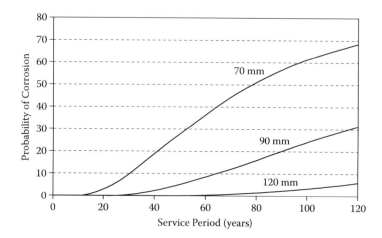

Figure 4.13 Effect of concrete cover on the probability of corrosion (type I concrete, 10°C).

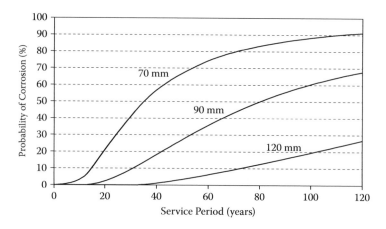

Figure 4.14 Effect of concrete cover on the probability of corrosion (type I concrete, 20°C).

than 120 years, respectively. Also, here, the effect of temperature is clearly demonstrated (Figure 4.14). For an increased annual temperature from 10 up to 20°C, even a concrete cover of 120 mm would no longer meet the required service period of 120 years before 10% probability of corrosion would be reached.

For the above calculations, it may be argued that any cover thickness of much more than 90 mm would not be very appropriate since the risk for unacceptable crack widths would be increased. While this effect to some extent could be mitigated by incorporating synthetic fibers in the concrete, a very thick concrete cover would also have some secondary effects, such as increased total dead load. An alternate solution to meet the required durability based on such a type of concrete, however, would be to replace the outer part of the rebar system with stainless steel, which would very effectively increase the cover thickness to the remaining black steel farther in. In this way, durability analyses can also be used as a design tool for quantifying how much of the traditional carbon steel needs to be replaced by stainless steel in order to meet the required safety level against corrosion.

4.6.3 Underwater infrastructure

4.6.3.1 Effect of concrete quality

In order to develop proper types of concrete for the future sea-spaced concrete structures in Singapore, a number of concrete mixtures based on various binder systems were produced, from which four different types were selected for further analyses (Teng and Gjørv, 2013). As a reference mixture, a concrete based on an ordinary portland cement (CEM I 42.5 R) with a maximum aggregate size of 20 mm was produced (Type A), while in the other mixtures, the portland cement was partly replaced by various combinations of a very finely ground granulated blast furnace slag (Blaine 870 m²/kg) and undensified silica fume, as shown in Table 4.7. By use of a high-range water-reducing admixture (HRWRA), the water/binder ratio for the mixtures varied from 0.25 to 0.28, but still all the fresh concrete mixtures were highly workable and stable, with properties similar to those of self-consolidating concrete.

Based on the obtained 28-day chloride diffusivities (D_{28}), a concrete cover of 70 mm (X_C), and estimated values for both the time dependence of the chloride diffusivity (α) and the critical chloride content (C_{CR}), as shown in Table 4.8, durability analyses were carried out. Also, for the other input parameters, the estimated values for environmental loading C_S (5.5; 1.35%), age at chloride loading t' (28 days), and temperature T (30°C) were kept constant.

Table 4.7 Concrete mixtures

Concrete Type	A	B	C	D
Water/binder ratio		0.28		0.25
Aggregate/binder ratio		3.35		2.85
Blast furnace slag replacement (%)	0	30	0	30
Silica fume replacement (%)	0	0	10	10
Total binder (kg/m³)	523	518	585	580
Coarse/fine aggregate ratio	1			
HRWRA/binder (%)	1	1	1.5	1.5
Aggregate by weight ratio	0.72	0.72	0.69	0.69

Table 4.8 Input parameters for analyzing the effect of concrete quality

Concrete	D_{28} (m²/s × 10⁻¹²)	α	C_{CR} (% by wt. of binder)	X_C (mm)
		Input parameter		
Type A	$N^a(7.9, 1.9)$	$N(0.4, 0.1)$	$N(0.4, 0.1)$	$N(70, 7)$
Type B	$N(1.0, 0.25)$	$N(0.5, 0.1)$		
Type C	$N(0.2, 0.05)$			
Type D	$N(0.1, 0.03)$			

ᵃ Normal distribution with average value and standard deviation.

As a result of the above durability analyses, all types of concrete apart from the reference concrete (Type A) would meet the required 150-year service period before 10% probability of corrosion would be reached. Although the reference concrete also had a very low water/binder ratio of 0.28, and thus would generally be considered very durable, Figure 4.15 clearly demonstrates that this concrete distinctly failed to meet the required durability. For the concrete of Type B, approximately 7% probability of corrosion would be reached, while for the two other types of concrete (Types C and D), the concrete had such a high resistance to chloride ingress that it was not possible to detect any probability of corrosion for a 150-year service period. Hence, both types of concrete would be very appropriate for possible future applications.

Again, it may be argued that the above durability analyses were only based on the obtained 28-day chloride diffusivities, while the chloride diffusivity of the various types of concrete with various types of binder system would develop somewhat differently. Therefore, here also some additional

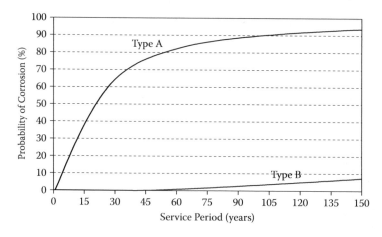

Figure 4.15 Effect of concrete cover on the probability of corrosion (Type A concrete).

durability analyses based on chloride diffusivities obtained after longer curing periods of up to 90 days were carried out, but none significantly changed the relative basis for comparing the effect of the various types of concrete.

4.6.3.2 Effect of concrete cover

In order to find out whether an increased concrete cover beyond 70 mm would also make it possible to use the reference concrete (mix A) and still meet the required durability, some further durability analyses based on 90 and 120 mm concrete cover were also carried out (Table 4.9). These analyses were only carried out for the reference concrete (Type A), while holding all the other input parameters from the previous analyses constant. From Figure 4.16 it can be seen, however, that even an increased concrete cover of up to 120 mm would not meet the required durability for this type

Table 4.9 Input parameters for analyzing the effect of concrete cover

Input parameter	Average value	Standard deviation	Comments
D_{28}	7.9	1.9	Chloride diffusivity (m²/s × 10⁻¹²)
α	0.5	0.1	Time dependence factor
C_{CR}	0.4	0.1	Critical chloride content (% by wt. of binder)
x_C	70	7	Nominal concrete cover (mm)
	90	7	
	120	7	

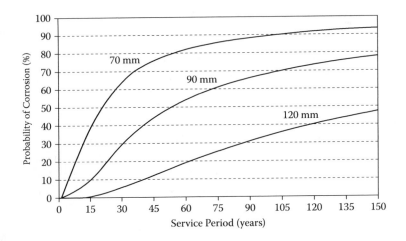

Figure 4.16 Effect of concrete quality on the probability of corrosion.

of concrete; the probability of corrosion would still reach approximately 50% for a 150-year service period.

From the above calculations it appears that all three types of concrete, Types B, C, and D, in combination with a concrete cover of 70 mm, would meet the required durability with a very good margin. For the concrete of Type B, approximately 7% probability of corrosion would be reached, while for the other two types of concrete (Types C and D), the 28-day chloride diffusivities were so low that it was not possible to detect any probability of corrosion for a 150-year service period (Teng and Gjørv, 2013).

If the above types of concrete (Types C and D) would also be combined with a somewhat increased concrete cover and a replacement of the black steel with stainless steel in the outer part of the rebar system, this would probably be a very good strategy for ensuring a very high durability of the given concrete structures in the given environment.

4.6.4 Evaluation and discussion of obtained results

For evaluation of the obtained results, it should be noted that the above calculations of corrosion probability are based on a number of assumptions and simplifications. Although diffusion is a most dominating transport mechanism through thick concrete covers in moist marine environments, only a very simple one-dimensional diffusion model for the calculation of chloride ingress rates is applied. As discussed in Chapter 3, the diffusion behavior of chloride ions in concrete is also a much more complex transport process than what can be described by Fick's second law of diffusion (Poulsen and Mejlbro, 2006; Zhang and Gjørv, 1996). Under more realistic conditions in the field, other transport mechanisms for chloride ingress than pure diffusion also exist. The characterization of the resistance of the concrete to chloride ingress is further based on a rapid migration type of testing, where the chloride ingress is very different from what takes place under more normal conditions in the field.

The above durability analyses are further based on a number of input parameters, for which there are a general lack of reliable data and information. In particular, this is true for the input parameters such as chloride loading and aging factor for the chloride diffusivity. Although a selection of these parameters should preferably be based on current experience from other similar concrete structures in similar environments, but such information may not necessarily exist or be available. Therefore, the selection of these parameters is mostly based on general experience. Age of the concrete at chloride loading and prevailing temperature are also other important parameters, proper values for which may also be somewhat difficult to select.

Based on all the above assumptions and simplifications, therefore, the obtained service periods with a probability for corrosion of less than 10%

should not be considered as real service periods for the given structures. The above case studies show, however, that the above durability analyses can be used as a basis for an engineering judgment of the most important parameters relevant for the durability, including the scatter and variability involved. The above case studies also clearly demonstrate how different chloride diffusivities, concrete covers, and temperatures may affect the durability of the given structures. Hence, a proper basis is obtained for comparing and selecting one of several technical solutions for a best possible durability of the given concrete structure in the given environment during the required service period.

REFERENCES

AASHTO. (2003). *AASHTO TP 64-03, Standard Method of Test for Prediction of Chloride Penetration in Hydraulic Cement Concrete by the Rapid Migration Procedure.* American Association of State Highway and Transportation Officials, Washington, DC.

Bamforth, P.B. (1999). The Derivation of Input Data for Modelling Chloride Ingress from Eight-Year Coastal Exposure Trials. *Magazine of Concrete Research*, 51(2), 87–96.

Bijen, J. (1998). *Blast Furnace Slag Cement for Durable Marine Structures.* VNC/BetonPrisma, Da's-Hertogenbosch, Netherlands.

Browne, R., et al. (1980). *Marine Durability Survey of the Tongue Sand Tower*, Concrete in the Ocean Program, CIRIA UEG Technical Report 5. Cement and Concrete Association, London.

CEB. (1992). *Durable Concrete Structures—Design Guide*, Comité Euro-International du Beton—CEB, Bulletin D'Information 183. Thomas Telford, London.

CEN. (2009). *Survey of National Requirements Used in Conjunction with EN 206-1:2000*, Technical Report CEN/TR 15868. CEN, Brussels.

ChlorTest. (2005). *WP 5 Report—Final Evaluation of Test Methods*, European Union Fifth Framework Program, Growth Project G6RD-CT-2002-00855: Resistance of Concrete to Chloride Ingress—From Laboratory Tests to In-Field Performance.

Collepardi, M., Marcialis, A., and Turriziani, R. (1970). Kinetics of Penetration of Chloride Ions in Concrete, *l'Industria Italiana del Cemento*, 4, 157–164.

Collepardi, M., Marcialis, A., and Turriziani, R. (1972). Penetration of Chloride Ions into Cement Pastes and Concretes. *Journal, American Ceramic Society*, 55(10), 534–535.

DuraCrete. (2000). *General Guidelines for Durability Design and Redesign, the European Union—Brite EuRam III*, Research Project BE95–1347: Probabilistic Performance Based Durability Design of Concrete Structures, Document R 15.

Engelund, S., and Sørensen, J.D. (1998). A Probabilistic Model for Chloride-Ingress and Initiation of Corrosion in Reinforced Concrete Structures. *Structural Safety*, 20, 69–89.

Ferreira, M. (2004). Probability Based Durability Design of Concrete Structures in Marine Environment, Doctoral Dissertation. Department of Civil Engineering, University of Minho, Guimarães, Portugal.

Ferreira, M., Årskog, V., Jalali, S., and Gjørv, O.E. (2004). Software for Probability-Based Durability Analysis of Concrete Structures. In *Proceedings, Fourth International Conference on Concrete under Severe Conditions—Environment and Loading*, vol. 1, ed. B.H. Oh, K. Sakai, O.E. Gjørv, and N. Banthia. Seoul National University and Korea Concrete Institute, Seoul, pp. 1015–1024.

FIB. (2006). *Model Code for Service Life Design*, FIB Bulletin 34. Federation International du Beton—FIB, Lausanne.

Fluge, F. (2001). Marine Chlorides—A Probabilistic Approach to Derive Provisions for EN 206-1. In *Proceedings, Third Workshop on Service Life Design of Concrete Structures—From Theory to Standardisation*. DuraNet, Tromsø, Norway, pp. 47–68.

Fredriksen, J.M., Sørensen, H.E., Andersen, A., and Klinghoffer, O. (1996). *The Effect of the w/c Ratio on Chloride Transport into Concrete*, HETEC Report 54. Danish Road Directorate, Copenhagen.

Gehlen, C. (2000). Probability-Based Service Life Calculations of Concrete Structures—Reliability Evaluation of Reinforcement Corrosion, Dissertation. RWTH-Aachen, Germany (in German).

Gehlen, C. (2007). Durability Design according to the New Model Code for Service Life Design. In *Proceedings, Fifth International Conference on Concrete Under Severe Conditions—Environment and Loading*, vol. 1, ed. F. Toutlemonde, K. Sakai, O.E. Gjørv, and N. Banthia. Laboratoire Central des Ponts et Chaussées, Paris, pp. 35–50.

Gehlen, C., and Schiessl, P. (1999). Probability-Based Durability Design for the Western Scheldt Tunnel. *Structural Concrete*, 1(2), 1–7.

Gjørv, O.E. (2002). Durability and Service Life of Concrete Structures. In *Proceedings, The First FIB Congress 2002*, session 8, vol. 6. Japan Prestressed Concrete Engineering Association, Tokyo, pp. 1–16.

Gjørv, O. E. (2003). Durability of Concrete Structures and Performance-Based Quality Control. In *Proceedings, International Conference on Performance of Construction Materials in the New Millennium*, ed. A.S. El-Dieb, M.M.R. Taha, and S.L. Lissel. Shams University, Cairo.

Gjørv, O.E. (2004). Durability Design and Construction Quality of Concrete Structures. In *Proceedings, Fourth International Conference on Concrete under Severe Conditions—Environment and Loading*, vol. 1, ed. B.H. Oh, K. Sakai, O.E. Gjørv, and N. Banthia. Seoul National University and Korea Concrete Institute, Seoul, pp. 44–55.

Hofsøy, A., Sørensen, S.I., and Markeset, G.A. (1999). *Experiences from Concrete Harbour Structures*, Report 2.2, Research Project Durable Concrete Structures. Norwegian Public Road Administration, Oslo (in Norwegian).

Hooton, R.D., Thomas, M.D.A., and Stanish, K. (2000). *Prediction of Chloride Penetration in Concrete*, Report FHWA-RD-00-142, U.S. Department of Transportation, Federal Highway Administration.

Kong, J.S., Ababneh, A.N., Frangopol, D.M., and Xi, Y. (2002). Reliability Analysis of Chloride Penetration in Saturated Concrete. *Probabilistic Engineering Mechanics*, 17(3), 305–315.

Lu, Z.-H., Zhao, Y.-G., and Yu, K. (2008). Stochastic Modeling of Corrosion Propagation for Service Life Prediction of Chloride Contaminated RC Structures. In *Proceedings, First International Symposium on Life-Cycle Civil Engineering*, ed. F. Biondini and D.M. Frangopol. Taylor & Francis Group, London, pp. 195–201.

Markeset, G. (2004). Service Life Design of Concrete Structures Viewed from an Owners Point of View. In *Proceedings, Seminar on Service Life Design of Concrete Structures*. Norwegian Concrete Association, Oslo, pp. 13.1–13.30 (in Norwegian).

Mangat, P.S., and Molloy, B.T. (1994). Prediction of Long-Term Chloride Concentration in Concrete. *Materials and Structures*, 27, 338–346.

McGee, R. (1999). Modelling of Durability Performance of Tasmanian Bridges. In *Proceedings, Eight International Conference on the Application of Statistics and Probability*, Sydney.

NAHE. (2004a). *Durable Concrete Structures—Part 1: Recommended Specifications for New Concrete Harbour Structures*. Norwegian Association for Harbour Engineers, TEKNA, Oslo (in Norwegian).

NAHE. (2004b). *Durable Concrete Structures—Part 2: Practical Guidelines for Durability Design and Concrete Quality Assurance*. Norwegian Association for Harbour Engineers, TEKNA, Oslo (in Norwegian).

NAHE. (2004c). *Durable Concrete Structures—Part 3: DURACON Software*. Norwegian Association for Harbour Engineers, TEKNA, Oslo.

Nilsson, L., Ngo M.H., and Gjørv, O.E. (1998). High-Performance Repair Materials for Concrete Structures in the Port of Gothenburg. In *Proceedings, Second International Conference on Concrete under Severe Conditions—Environment and Loading*, vol. 2, ed. O.E. Gjørv, K. Sakai, and N. Banthia. E & FN Spon, London, pp. 1193–1198.

NORDTEST. (1989). *NT Build 355: Concrete, Repairing Materials and Protective Coating: Diffusion Cell Method, Chloride Permeability*. NORDTEST, Espoo, Finland.

NORDTEST. (1995). *NT Build 443: Concrete, Hardened: Accelerated Chloride Penetration*. NORDTEST, Espoo, Finland.

NORDTEST. (1999). *NT Build 492: Concrete, Mortar and Cement Based Repair Materials: Chloride Migration Coefficient from Non-Steady State Migration Experiments*. NORDTEST, Espoo, Finland.

NTU. (2011). NRF Research Program *Underwater Infrastructure and Underwater City of the Future*. Nanyang Technological University, Singapore.

PIANC/NAHE. (2009a). *Durable Concrete Structures—Part 1: Recommended Specifications for New Concrete Harbour Structures*, 3rd ed. Norwegian Association for Harbour Engineers, TEKNA, Oslo (in Norwegian).

PIANC/NAHE. (2009b). *Durable Concrete Structures—Part 2: Practical Guidelines for Durability Design and Concrete Quality Assurance*, 3rd ed. Norwegian Association for Harbour Engineers, TEKNA, Oslo (in Norwegian).

PIANC/NAHE. (2009c). *Durable Concrete Structures—Part 3: DURACON Software*, 3rd ed. Norwegian Association for Harbour Engineers, TEKNA, Oslo.

Poulsen, E., and Mejlbro, L. (2006). *Diffusion of Chlorides in Concrete—Theory and Application*. Taylor & Francis, London.

Schiessl, P., and Lay, S. (2005). Influence of Concrete Composition. In *Corrosion in Reinforced Concrete Structures*, ed. H. Böhni. Woodhead Publishing, Cambridge, UK, pp. 91–134.

Siemes, A.J.M., and Rostam, S. (1996). Durability Safety and Serviceability—A Performance Based Design. In *Proceedings, IABSE Colloquium on Basis of Design and Actions on Structures*, Delft, Netherlands.

Standard Norway. (2004). *NS 3490: Design of Structures—Requirements to Reliability*. Standard Norway, Oslo (in Norwegian).

Stewart, M.G., and Rosowsky, D.V. (1998). Structural Safety and Serviceability of Concrete Bridges Subject to Corrosion. *Journal of Infrastructure Systems*, 4(4), 146–155.

Takewaka, K., and Mastumoto, S. (1988). Quality and Cover Thickness of Concrete Based on the Estimation of Chloride Penetration in Marine Environments. In *Proceedings, Second International Conference on Concrete in Marine Environment*, ACI SP 109, ed. V.M. Malhotra, pp. 381–400.

Tang, L. (1996a). Electrically Accelerated Methods for Determining Chloride Diffusivity in Concrete. *Magazine of Concrete Research*, 48(176), 173–179.

Tang, L. (1996b). *Chloride Transport in Concrete—Measurement and Prediction*, Publication P-96:6. Department of Building Materials, Chalmers University of Technology, Gothenburg.

Tang, L., and Gulikers, J. (2007). On the Mathematics of Time-Dependent Apparent Chloride Diffusion Coefficient in Concrete. *Cement and Concrete Research*, 37(4), 589–595.

Tang, L., Nilsson, L.-O., and Basher, P.A.M. (2012). *Resistance of Concrete to Chloride Ingress*. Spon Press, London.

Tang, L., and Sørensen, H.E. (2001). Precision of the Nordic Test Methods for Measuring the Chloride Diffusion/Migration Coefficients of Concrete. *Materials and Structures*, 34, 479–485.

Teng, S., and Gjørv, O.E. (2013). Concrete Infrastructures for the Underwater City of the Future. In *Proceedings, Seventh International Conference on Concrete under Severe Conditions—Environment and Loading*, ed. Z.J. Li, W. Sun, C.W. Miao, K. Sakai, O.E. Gjørv, and N. Banthia. RILEM, Bagneux, pp. 1372–1385.

Thomas, M.D.A., and Bamforth, P.B. (1999). Modelling Chloride Diffusion in Concrete—Effect of Fly Ash and Slag. *Cement and Concrete Research*, 29, 487–495.

Thomas, M.D.A., Bremner, T., and Scott, A.C.N. (2011). Actual and Modeled Performance in a Tidal Zone. *Concrete International*, 33(11), 23–28.

Thomas, M.D.A., Shehata, M.H., Shashiprakash, S.G., Hopkins, D.S., and Cail, K. (1999). Use of Ternary Cementitious Systems Containing Silica Fume an Fly Ash in Concrete. *Cement and Concrete Research*, 29, 1207–1214.

Tong, L., and Gjørv, O.E. (2001). Chloride Diffusivity Based on Migration Testing. *Cement and Concrete Research*, 31, 973–982.

Tuutti, K. (1982). *Corrosion of Steel in Concrete*, Report 4. Swedish Cement and Concrete Institute, Stockholm.

Zhang, T., and Gjørv, O.E. (1996). Diffusion Behavior of Chloride Ions in Concrete. *Cement and Concrete Research*, 26, 907–917.

Chapter 5

Additional strategies and protective measures

5.1 GENERAL

For all major concrete infrastructures, a service period of 100 years or more should normally be required before the probability of corrosion exceeds 10%. However, since the calculations of corrosion probability gradually become less reliable for increasing service periods beyond 100 years, as discussed in Chapter 4, the corrosion probability should be kept as low as possible, not exceeding 10% for a service period of up to 150 years, but in addition, some special protective measures should also preferably be applied. Since any calculations of corrosion probability for service periods of more than 150 years are no longer considered valid, the corrosion probability for such long service periods should still be kept as low as possible, not exceeding 10% for a 150-year service period, but in addition, some special protective measures should always be applied in order to further increase and ensure the durability.

For concrete construction work in marine environments, there may also be a risk for early-age exposure during concrete construction before the concrete has gained sufficient maturity and density, as shown in Chapter 2. Also for such a case, some special precautions or protective measures should be considered. Since the special protective measure may have implications both for the economy of the project and for the future operation of the structure, such measures should always be discussed with the owner of the structure before the special strategy and protective measure are selected.

For major concrete infrastructures with a required service period of more than 100 years, a partial replacement of the black steel reinforcement with stainless steel has proved to be a most efficient protective measure for ensuring a high and reliable long-term performance (Chapter 2). Since concrete structures typically show a high scatter and variability of achieved construction quality, and any weaknesses and deficiencies soon will be revealed, a partial use of stainless steel has also proved to be a very simple and robust protective measure. For concrete structures in chloride-containing environments, therefore, stainless steel reinforcement is described in more detail, but also a number of other protective measures exist that are briefly outlined and discussed in the following.

5.2 STAINLESS STEEL REINFORCEMENT

For a long time, stainless steel reinforcement has proved to be a very efficient way of enhancing durability and service life of concrete structures, even in the most severe marine environments with elevated temperatures. As was shown in Chapter 2, reinforcement based on stainless steel (AISI 304/W.1.4301) was applied in a concrete pier on the Yucatán Coast in Mexico already in 1937, and experience later on has shown that the additional cost of this protective measure proved to be an extremely good investment for the owner. Traditionally, the costs of stainless steel have been so high that it has not normally been considered viable for ordinary concrete structures. During recent years, however, new experience has shown that a more selective use of stainless steel in the most critical parts of the structure can be very attractive for enhancing durability and service life of concrete structures compared to those of other special protective measures (Knudsen et al., 1998; Materen and Poulsen-Tralla, 2001; Knudsen and Goltermann, 2004).

For many years, it was believed that a galvanic coupling between reinforcing bars based on stainless steel and carbon steel would represent a potential corrosion problem. Both extensive experimental investigations and practical experience have demonstrated, however, that a partial use of stainless steel in coupling with carbon steel in concrete does not increase the risk of corrosion (Bertolini et al., 2004). As a consequence, a partial replacement of the carbon steel by stainless steel in the most exposed parts of the structure has proved to be very good, from both a protective and a cost-effective point of view.

There are many different qualities of stainless steel reinforcement on the market, but depending on both the chemical composition and the microstructure of the steel, there are basically the following three groups:

- Ferritic steel
- Austenitic steel
- Austenitic–ferritic steel (duplex)

The corrosion resistance required for chloride-containing environments mainly depends on the alloying elements, such as chromium, nickel, molybdenum, and nitrogen, but the microstructure is also important. A classification of the various types of stainless steel is given in both the European Standard EN 10088-1 (CEN, 1995) and the U.S. Standard AISI.

For stainless steel in concrete, however, it should be noted that the corrosion resistance also depends on a number of other factors, such as the potential of the embedded steel, which can vary depending on the oxygen availability (Bertolini and Gastaldi, 2011). Thus, the applicability of the various types of stainless steel is increased when the free corrosion potential is reduced, such as in water-saturated concrete. It should further be noted that both the presence of welding scale and poor surface finishing of the steel will reduce the critical chloride content (Pediferri, 2006). When the surfaces of austenitic types of steel such as 1.4307 and 1.4404 were covered by a welding scale, reduced critical chloride contents (about 3.5%) were observed (Sørensen et al., 1990; Pediferri et al., 1998; Bertolini et al., 1998). Elevated temperatures will also reduce the critical chloride content for the given type of steel (Bertolini and Gastaldi, 2011). Recent experience has shown that the so-called pitting resistance equivalent number (PREN) is not very applicable for assessment of the corrosion resistance of various types of stainless steel in concrete (Bertolini, 2012).

In order to reduce the cost of stainless steel reinforcement, there has been an increased focus in recent years to reduce the cost of the raw materials. In particular, the cost of nickel typically shows considerable fluctuation, which for periods has shown a doubling of its cost (LME). As a consequence, lean types of duplex steel with low contents of nickel and also molybdenum have shown to be more cost-effective. In Table 5.1, approximate chemical composition and designation of some typical grades of stainless steel used for reinforcing bars are shown. In this table, the notation L indicates that the steel has a low carbon content and is thus weldable.

For assessment of the corrosion resistance of both carbon steel and stainless steel in concrete, there is currently a lack of proper test methods and testing procedures. In the literature, studies and results on the corrosion resistance and data on critical chloride contents for various types of steel

Table 5.1 Approximate chemical composition and designation of some typical grades of stainless steel used for reinforcing bars

	Designation			Approximate chemical composition (% by mass)			
Grade	AISI	EN 10088-1	Microstructure	Cr	Ni	Mo	Other elements
304L	304L	1.4307	Austenitic	17.5–19.5	8–10	—	—
316L	316L	1.4404	Austenitic	16.5–18.5	10–13	2–2.5	—
22-05	318	1.4462	Duplex	21–23	4.5–6.5	2.5–3.5	N
23-04	—	1.4362	Duplex	22–24	3.5–5.5	0.1–0.6	N
21-01	—	1.4162	Duplex	21–22	1.4–1.7	0.1–0.8	Mn, N

Source: Bertolini, L., and Gastaldi, M., *Materials and Corrosion*, 62, 120–129, 2011.

are also scarce, and the results available are partly based on investigations of steel in solutions and partly based on embedded steel in concrete. Since such a different approach to the testing may give different results, only tests carried out on embedded steel in concrete should be considered suitable for evaluating the corrosion resistance of the steel (Bertolini and Gastaldi, 2011).

While a chloride threshold value of 0.4% by weight of cement is often assumed for traditional carbon steel in concrete, indicative threshold values for austenitic stainless steel of types 1.4301 and 1.4401 may vary from 3.5 to 5% and 3.5 to 8%, respectively (Bertolini et al., 2004). For the duplex types of stainless steel, such as 1.4362 and 1.4462, the values may also vary from 3.5 to 5% and 3.5 to 8%, respectively. The duplex steel 1.4462 has further shown very good corrosion resistance at elevated temperatures of up to 40°C (Bertolini and Gastaldi, 2011; The Concrete Society, 1998).

Although some of the above types of stainless steel may be safely used for chloride contents of up to 5 or even 8%, such high chloride contents are rarely ever reached in the vicinity of steel embedded in concrete. Hence, a partial replacement of the carbon steel with lean types of duplex stainless steel can substantially enhance the durability and service life of the structure. From tests on lean duplex types of steel such as 1.4162 or 1.4362 in concrete specimens exposed at 20°C and 90% relative humidity (RH), chloride thresholds in the order of 2.5 and 3% by weight of cement are reported (Bertolini and Gastaldi, 2011).

Reinforcement in the form of stainless steel-clad carbon steel is also available on the market (Rasheeduzzafar et al., 1992; Clemeña, 2002). Although a stainless steel cladding may also give an effective protection, possible defects in the cladding during bending may reduce the protective efficiency (Clemeña and Virmani, 2004).

In order to quantify how much of the traditional carbon steel needs to be replaced by stainless steel in the most vulnerable parts of the concrete structure, it was already shown how durability analyses can be an effective design tool, as previously discussed in Chapter 4. In this case, the increased depth of concrete cover to the more vulnerable carbon steel farther below the concrete surface is used as an input parameter to the durability analysis. If the stainless steel also meets the requirements for mechanical properties, bond strength and bending, etc., all traditional procedures for both structural design and construction can still be followed.

By replacing up to 40% of the traditional carbon steel reinforcement by stainless steel in the most vulnerable parts of a typical concrete harbor structure, calculations showed that the total cost of the structure did not increase by more than about 5% (Isaksen, 2004). In the United States, the Oregon State Department of Transportation has already, for many years, specified a partial use of stainless steel reinforcement for new concrete bridge construction along the coastline of Oregon (Cramer et al., 2002). This specification is based on requirements to both corrosion threshold value and corrosion rate, as well as yield strength, and the specification is allowable in the sense that it allows the contractor to choose between different types of stainless steel. In addition to the specification of a high-performance type of concrete with very high resistance to chloride ingress, stainless steel is being specified for both deck beams and precast prestressed girders of the bridges. In spite of the high additional costs of stainless steel, the total project costs for three concrete coastal bridge projects only increased by about 10% compared to the equivalent quantity of traditional carbon steel (Table 5.2). At a minimum, it was anticipated that the use of stainless steel in these bridges would double the bridge life while cutting cumulative costs relative to conventional steel reinforcement by 50% over the specified service life of 120+ years.

Table 5.2 Materials costs for three concrete coastal bridges in Oregon using stainless steel bars in the most exposed parts of the bridges

Project	Brush Creek (1998)	Smith River (1999)	Haynes Inlet (2003)
Stainless Steel Bar			
Uses	Deck beams	Precast, prestressed girders[a]	Deck beams
Alloy	316N	316N	316LN
Yield strength (MPa)	414	414	517
Unit price ($/kg)	7.88	262.47/ girder-meter	5.02
Quantity (kg)	42,270	2713 girder-meters	320,000
Total stainless cost[b] ($)	333,660	712,080	1,610,000
Equivalent black iron bar cost[b] ($)	107,790	Not available	486,400
Black Iron Bar			
Unit price ($/kg)	2.55	Not available	1.52
Quantity (kg)	69,550	Not available	600,000
Total black iron bar cost[b] ($)	187,020	390,900	900,000
Project Summary			
Total project cost[b] ($)	2,259,380	8,565,080	11,055,400
Stainless cost as % of project cost	14.8%	8.3%	14.5%
Stainless cost premium over black iron bar as % of project cost	10.0%	Not available	10.2%

Source: Cramer, S. D. et al., Cement and Concrete Composites, 24, 101–117, 2002.

[a] Reinforcing bar cast in girders.
[b] 1999 U.S. dollars.

In addition to being a cost-effective protective measure for concrete structures in severe environments, a proper utilization of stainless steel has also proved to be a very simple and robust strategy for achieving a more controlled and enhanced durability and service life of the structures. For further information about the selection and use of stainless steel in concrete, reference is made to both the catalogs from the stainless steel producers and the more specialized literature referred to in the list of references.

5.3 OTHER PROTECTIVE MEASURES

5.3.1 Nonmetallic reinforcement

In recent years, applications of fiber-reinforced polymer (FRP) composites as reinforcement for concrete structures have been growing rapidly. Although most of the current experience and durability data on FRP composite installations come from the aerospace, marine, and corrosion resistance industries, FRP composites have been used as a construction material since the mid-1950s (ACMA MDA, 2006). A major development of FRP for civil engineering has been the application of externally bonded FRP for rehabilitation and retrofitting of existing concrete structures. During the late 1970s and early 1980s, however, a variety of new applications of composite reinforcing products were demonstrated, and already in 1986, the world's first highway bridge using composite tendons was built in Germany. In recent years, nonmetallic reinforcement based on FRP composites has found a wide range of applications, and current experience demonstrates that such reinforcement systems have great potential for concrete structures in severe environments (Newhook and Mufti, 1996; Newhook et al., 2000; Tan, 2003; Serancio, 2004; Newhook, 2006).

Since FRP composites may show very high mechanical properties (Table 5.3), such products also represent a viable alternative to conventional steel tendons for prestressed and post-tensioned concrete. For a long time, suitable anchorage systems were a problem, but in recent years, new types of anchorage systems based on FRP tendons have been developed for practical applications of prestressed systems (Gaubinger et al., 2002).

Until recently, glass fibers were the most predominant type of reinforcing fiber, mostly in the form of E-glass formulation, but also in the form of alkali resistance glass (AR glass). Due to the rapid increase in production capacity and new production methods, however, the costs of both carbon and aramid fibers have also made such fibers more easily available and attractive. In recent years, basalt fibers with improved properties compared to those of glass fibers have also been introduced (ReforceTech, 2013). These fibers do not have any durability problems in the highly alkaline environment of concrete. Also, the cost of these fibers is equivalent to that

Table 5.3 Mechanical properties of advanced composite fibers

	Armid fiber (Twaron HM)	Glass fiber (E-glass)	Basalt fiber	Carbon fiber (HT)	CFRP wire	Steel strand (St. 1570/1770)
Tensile strength (MPa)	2600	2300	3200	3500/7000	2800	>1770
Young's modulus (GPa)	125	74	90	230/650	160	205
Ultimate strain (%)	2.3	3.3	3.0	0.6/2.4	1.6	7
Density (g/cm³)	1.45	2.54	2.6	1.8	1.5	7.85

Source: Noisternig, F. et al., *Development of CFK Prestressed Elements*, Seminarband Kreative Ingenieurleistungen, Darmstadt–Wien, 1998 (in German); ReforceTech, Basalt Fiber Reinforcement Technology, 2013, http://reforcetech.com.

of glass fibers, and approximately 1/10 that of carbon fibers. Embedded in a proper type of a polymer-based matrix, all the above types of fibers are currently commercially available in various qualities and dimensions as reinforcing bars for concrete reinforcement.

In principle, the fundamental design methodology for FRP products is similar to that of conventional steel-reinforced concrete. Cross-sectional equilibrium, strain compatibility, and constitutive material behavior form the basis of all approaches to the structural design of reinforced concrete structures, regardless of the reinforcing material. For a proper structural design, however, the nonductile and anisotropic natures of FRP reinforcing products need to be specially addressed, but this appears to be properly taken into account in current guidelines and recommendations for structural design (CSA, 2003; ACI, 2007; FIB, 2007).

5.3.2 Concrete surface protection

As discussed in Chapter 2, those offshore concrete platforms that received a thick epoxy coating applied to the concrete surface during concrete construction have been very effectively protected from chloride ingress (FIP, 1996; Årstein et al., 1998). As this surface coating was continuously applied to the concrete surface during slip forming, when the young concrete still had an underpressure and suction ability, a very good bond between the concrete substrate and the coating was achieved.

In recent years, a number of new surface protection products have been introduced for either retarding or preventing the ingress of chlorides into concrete structures. In principle, the effect of such products may be either to make the concrete surface less permeable to chloride ingress or to reduce the moisture content in the concrete, although many products combine these effects.

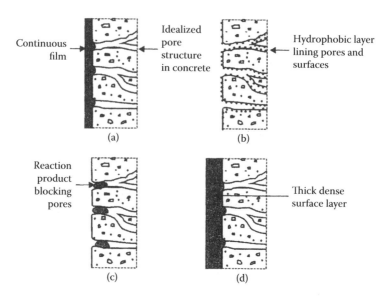

Figure 5.1 Schematic representation of different types of concrete surface protection: (a) organic coating, (b) pore-lining treatments, (c) pore-blocking treatments, and (d) thick cementitious coating, shotcrete, or rendering. (From Bijen, J. M., *Heron*, 34(2), 2–82, 1989.)

Different types of surface protection products may be grouped into the following four classes, as schematically shown in Figure 5.1: (a) organic coatings that form a continuous film on the concrete surface, (b) hydrophobic treatments that line the interior surface of the concrete pores, (c) treatments that fill the capillary pores, and (d) a thick and dense cementitious layer. While surface treatments open to water vapor may have a longer service life with no or slower loss of bond, the effectiveness of such surface treatments may generally be less than that of dense protective systems. It appears, therefore, to be a trade-off between good barrier properties and good long-term effects of the surface treatment.

In recent years, surface hydrophobation of concrete structures has been widely adopted as a protective measure for concrete structures in severe environments. Since the protective efficiency of such a polymer-based surface treatment may be reduced over time due to weathering, most types of such treatments need to be reapplied from time to time in order to ensure proper long-term protection. If the surface hydrophobation is applied at a later stage, where the chlorides have already reached a certain depth, a further chloride ingress may still take place for some time due to a redistribution of the chloride content (Arntsen, 2001; Årskog et al., 2004). However, in severe marine environments, a surface hydrophobation may primarily be applied during concrete construction in order to protect the concrete

Table 5.4 Effect of temperature and moisture
conditions on penetration depth
of hydrophobic agent

Code	Penetration depth (mm)
T-20-50	9.6 ± 1.2
T-20-100	2.0 ± 2.8
T-05-95	2.8 ± 0.3
T-05-100	<0.1

Source: Liu, G. et al., Effect of Surface Hydrophobation for Protection of Early Age Concrete against Chloride Penetration, in *Proceedings, Fourth International Conference on Water Repellent Treatment of Building Materials*, ed. J. Silfwerbrand, Aedificatio Publishers, Freiburg, Germany, 2005, pp. 93–104.

against early-age exposure before the concrete has gained sufficient maturity and density.

In order to investigate the protective effect of a surface hydrophobation against early-age chloride exposure, an experimental program based on a concrete with a water/cement ratio of 0.45 was carried out (Liu et al., 2005). After seven days of concrete curing at a temperature of either 20 or 5°C and an RH of 50, 95, or 100%, concrete specimens were surface treated with a hydrophobic gel, consisting mainly of an isobutyl triethoxy type of silane. Before exposure, a penetration depth of the hydrophobic agent, as shown in Table 5.4, was observed, and after six weeks of intermittent spraying and drying to a 3% NaCl solution, the protective efficiency was evaluated on the basis of depth of chloride penetration (C_x), surface chloride concentration (C_s), and apparent chloride diffusivity (D_a), calculated on the basis of Fick's second law of diffusion. In addition, the chloride penetration rate (V) was also calculated as the total amount of penetrated chlorides divided by the area of exposed surface and time of exposure (g/m²·s).

For the curing conditions at 20°C and 50% RH, it can be seen from Table 5.5 that the surface treatment reduced the depth of chloride penetration, apparent chloride diffusivity, and chloride penetration rate from 7.8 to 1.6 mm, from 6.5 to 0.3 × 10⁻¹² m²/s and from 3.0 to 0.9 × 10⁻⁵ g/m²·s, respectively. For increased relative humidity to 100%, however, a depth of chloride penetration, apparent chloride diffusivity, and chloride penetration rate of 5.1 mm, 4.4 × 10⁻¹² m²/s, and 1.7 × 10⁻⁵ g/m²·s, respectively, were observed. For a combination of increased humidity to 100% RH and reduced temperature to 5°C, the corresponding numbers were 7.7 mm, 8.3 × 10⁻¹² m²/s, and 1.7 × 10⁻⁵ g/m²·s, respectively. Although the surface chloride concentration did not reflect the efficiency of the surface treatment, there was a good correlation between the penetration depth of the hydrophobic agent (Table 5.5) and the protective efficiency of the surface treatment. As can be seen from both Table 5.5 and Figure 5.2, however,

Table 5.5 Chloride penetration in untreated (U) and treated (T) concrete surface after early-age exposure

Code	Penetration depth (C_x) (mm)	Surface concentration (C_s) (% concrete weight)	Apparent chloride diffusion coefficient (D_a) (10^{-12} m²/s)	Chloride penetration rate (V) (10^{-5} g/m² s)
U-20-50	7.8 ± 1.0	0.64	6.50	2.96
T-20-50	1.6 ± 0.5	0.41 ± 0.02	0.29 ± 0.03	0.87 ± 0.01
U-20-100	6.9 ± 1.0	0.35	6.10	1.50
T-20-100	5.1 ± 1.1	0.40 ± 0.10	4.42 ± 0.74	1.68 ± 0.30
U-05-95	10.0 ± 0.8	0.62	9.80	3.33
T-05-95	5.4 ± 1.4	0.19 ± 0.02	12.80 ± 2.97	1.16 ± 0.02
U-05-100	9.1 ± 0.5	0.61	6.74	3.06
T-05-100	7.7 ± 0.1	0.34 ± 0.02	8.29 ± 0.05	1.72 ± 0.07

Source: Liu, G. et al., Effect of Surface Hydrophobation for Protection of Early Age Concrete against Chloride Penetration, in *Proceedings, Fourth International Conference on Water Repellent Treatment of Building Materials*, ed. J. Silfwerbrand, Aedificatio Publishers, Freiburg, Germany, 2005, pp. 93–104.

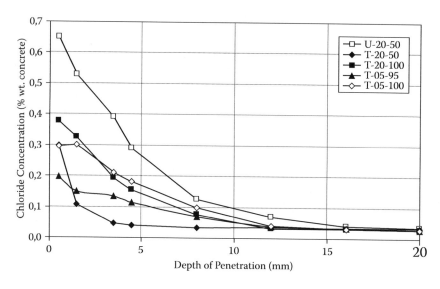

Figure 5.2 Protective effect of a surface hydrophobation against early-age chloride exposure. (From Liu, G. et al., Effect of Surface Hydrophobation for Protection of Early Age Concrete against Chloride Penetration, in *Proceedings, Fourth International Conference on Water Repellent Treatment of Building Materials*, ed. J. Silfwerbrand, Aedificatio Publishers, Freiburg, Germany, 2005, pp. 93–104.)

the protective efficiency of the surface treatment was not very good on the moist concrete substrate. For the moist substrate, a low temperature was also very important for the efficiency of the surface treatment.

For many concrete structures in severe environments, the moisture content in the surface layer of the concrete may be quite high (Chapter 2). Therefore, in order to investigate the protective efficiency of a surface hydrophobation under more realistic conditions in the field, some tests on a new concrete harbor structure shortly after concrete construction were carried out (Liu, 2006). The surface treatments included two types of gel-based products, one of which consisted mainly of an isooctyl trietoxy type of silan (T-A), while the other was based on an isobutyl trietoxy silan (T-B). Both products were partly applied in one layer with achieved thickness of approximately 0.25 mm, and in two layers with achieved thickness of approximately 0.5 mm. The moisture content in the concrete substrate was so high that none of the surface treatments gave any observed depth for penetration of the hydrophobic agent. After four years of exposure to splashing of seawater, however, it can be seen from Figure 5.3 that all the applied protective systems had provided about the same good protection against chloride penetration. Compared to the observed chloride penetration in the unprotected parts of the concrete deck, the surface hydrophobation had reduced the surface chloride concentration, apparent chloride diffusivity, and chloride penetration rate by factors of 0.5, 0.7, and 0.5, respectively.

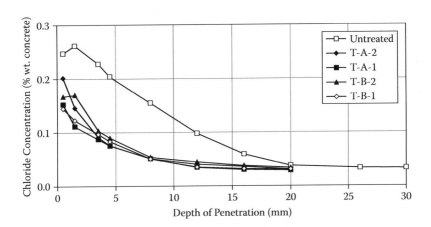

Figure 5.3 Effect of hydrophobic surface treatment for the early-age protection of a concrete harbor structure against chloride penetration. (From Liu, G., Control of Chloride Penetration into Concrete Structures at Early Age, Dr.Ing. Thesis 2006:46, Department of Structural Engineering, Norwegian University of Science and Technology—NTNU, Trondheim, Norway, 2006.)

In the literature, extensive experience with various types of protective surface systems applied to concrete structures in severe environments has been reported, good reviews of which are given in COST Action 521 (COST, 2003), Bertolini et al. (2004), and Raupach and Rößler (2005). The proceedings from the international conferences on water-repellent treatment of building materials also reflect much of the current knowledge and experience with this type of protective surface system (Silfwerbrand, 2005; De Clereq and Charola, 2008).

5.3.3 Concrete hydrophobation

By also using silane-based hydrophobic agents as an admixture to the fresh concrete, it is possible to make the whole mass of the concrete hydrophobic. Such an approach may be appropriate either for certain critical parts of the structure or for thin concrete elements, such as precast formwork elements. If the concrete is allowed to properly dry out before exposure, investigations have shown that this may be a viable alternative to the traditional surface hydrophobation of concrete (Årskog et al., 2011).

5.3.4 Cathodic prevention

For concrete structures that already have ongoing chloride-induced corrosion, extensive experience has shown that repairs based on a cathodic protection (CP) system probably are the most effective way of getting such corrosion under control. For the design of a proper CP system, however, the challenge is always to provide the necessary protective current in a reliable, controllable, and durable way (Bertolini et al., 2008). For such a protective measure to be effective, a proper electrical continuity within the rebar system must therefore be provided. According to the European standard for cathodic protection of concrete structures (CEN, 2000), the electrical resistance between any two points in the rebar system shall not exceed 1 ohm. Providing such electrical continuity during the repair may be technically much more difficult and substantially more costly than providing such continuity during construction of the new structure. Specifying a proper electrical continuity within the rebar system as part of the durability design may therefore be a proper strategy for future repairs based on cathodic protection. For new concrete structures, an even better strategy might be to not only specify provisions for future repairs based on a cathodic protection system, but also install the cathodic protection system right from the beginning, or install it a later stage of operation, before the chlorides reach the embedded steel and start any corrosion. By denoting this approach for cathodic prevention, this technique was suggested and introduced by Pediferri in the late 1980s (Pediferri, 1992). Later on, cathodic prevention

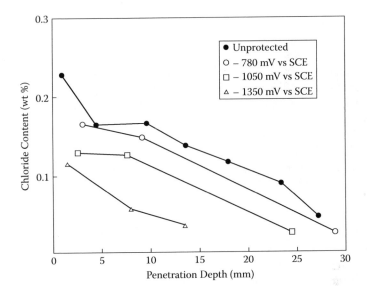

Figure 5.4 Effect of protection potential on rate of chloride diffusion through sub-merged concrete. (From Gjørv, O. E., and Vennesland, Ø., Evaluation and Control of Steel Corrosion in Offshore Concrete Structures, in *Proceedings, The Katharine and Bryant Mather International Conference*, vol. 2, ed. J. Scanlon, ACI SP-1, 1987, pp. 1575–1602.)

was successfully applied to a number of concrete structures (Broomfield, 1997; Bertolini, 2000; COST, 2003).

In principle, cathodic prevention also increases the chloride threshold in the concrete as the electrochemical potential of the embedded steel is decreased. According to Bertolini et al. (2004), very low current densities of less than 2 mA/m^2 are needed in order to bring the potential to values that will suppress initiation of pitting corrosion even at high chloride concentrations at the steel surface. For concrete structures submerged in seawater, cathodic prevention may also effectively counteract the rate of chloride diffusion through the concrete cover, as demonstrated in Figure 5.4.

In order to compare cathodic prevention with other protective measures, it may be difficult to estimate the long-term performance and service life of the protective system. In addition to the initial costs of installation, the costs for operation and maintenance of the electrode system, wires, and sensitive electronic equipment must also be properly taken into account (Broomfield, 1997; COST, 2003).

Although there is not much experience reported on the long-term performance of cathodic prevention systems, recent investigations of 105 installations of cathodic protection systems in the Netherlands have been carried out, of which 50 installations had been operating more than 10 years or longer (Polder et al., 2012). The service period without any major intervention varied between 10 and 20 years, but the need for some extent of intervention typically increased for increased age of the installation. In order to generalize the observations, a survival analysis was carried out showing that there was a 10% probability for necessary intervention after seven years or less, and a 50% probability for necessary intervention after a service period of 15 years or less. As a basis for these results, however, it was emphasized that a proper maintenance of the cathodic protection systems had been carried out, which involved electrical testing for depolarization at least twice a year and visual inspection once a year, as described by the European CP standard (CEN, 2000); such maintenance is normally a part of the maintenance contract between the owner and the CP contractor.

If only provisions for future installation of a cathodic prevention system for the structure are specified, it should be noted that such a system has to be installed at a stage before the first chlorides have reached embedded steel and caused any corrosion. Hence, a very close control and following up of the chloride ingress during operation of the structure must be carried out. Since all concrete structures typically show a high scatter and variability of achieved construction quality, however, this may represent a great challenge for the owner during operation of the structure (Chapter 8). Also, larger parts of the concrete structures may later on not necessarily be easily accessible for neither control of chloride ingress nor any installation of cathodic prevention.

Although current experience with both cathodic prevention and cathodic protection is quite good, it should be noted that such systems are based on sensitive electronic equipment in combination with a number of different types of components typically exposed to an aggressive environment. Therefore, regular control and maintenance by qualified personnel must be carried out; over time, power units may stop working, anode-copper connections may corrode, reference electrodes may fail, and anode materials or primary anodes may degrade.

5.3.5 Corrosion inhibitors

Corrosion inhibitors for both prevention and delay of corrosion initiation have been on the market for a long time. A number of different inhibitors for addition to the fresh concrete exist, but from one type of product to another, the protecting mechanisms may be very different (Büchler, 2005). Of the various types of product, calcium nitrite is probably the most extensively tested and widely applied so far. Extensive investigations have shown that a proper addition of this inhibitor is capable of both preventing corrosion and decreasing the corrosion rate (Hinatsu et al., 1990). In order to be efficient, however, a critical ratio in the range of 0.5 to 1.0 between nitrite and chloride has to be present (Andrade, 1986; Gaidis and Rosenberg,

1987). Also, since corrosion activation will consume the substance over time, while the concentration of chlorides will increase, it may be difficult to predict and guarantee the long-term effect of such a protective measure (Hinatsu et al., 1990). If the concentration of the calcium nitrite over time becomes too low, corrosion acceleration may take place (Nürnberger, 1988; Ngala et al., 2002).

5.3.6 Structural design

Already at an early stage of field experience with concrete structures in marine environments, it was shown in Chapter 2 that open concrete harbor structures with a flat slab type of deck typically performed much better than structures having a beam and slab type of deck. For the structures with a beam and slab type of deck, the deck slab sections also typically showed a much better performance than the deck beams.

For a long time it was assumed that the different behaviors of these different types of deck were due to an easier and better placing and compaction of the fresh concrete in the deck slabs than that of deep and narrow beams and girders. In Norwegian concrete harbor construction, the practical consequence of this experience was drawn already in the early 1930s, when the first open concrete harbor structure with a flat slab type of deck was introduced (Figure 2.13). From then on, a number of new concrete harbor structures with a flat slab type of deck were constructed, all of which showed a much better long-term performance than those structures having a beam and slab type of deck.

Since the new structural design based on flat slab decks was often more expensive, however, the slab and beam type of deck was gradually introduced again. It was assumed that if only the beams were made shallower and wider, it would be equally easy to place the fresh concrete in the deck beams. After some time, however, even the shallower and wider deck beams again showed early steel corrosion, while the flat slab type of deck still performed very well.

What was not known in the earlier days was that the more exposed deck beams of the open concrete harbor structures would always absorb and accumulate more chlorides than the less exposed deck slab sections in between. As a consequence, the steel in the more exposed deck beams would soon depassivate, and hence develop anodic areas, while the less exposed slab sections in between would act as catchments areas for oxygen, and hence form cathodic areas. In this way, a complex system of galvanic cell activities along the embedded steel in the concrete deck would develop with accelerated corrosion in the deck beams, typically functioning as sacrificial anodes, and thus cathodically protecting the slab sections in between.

Therefore, for structural design of concrete structures in marine environments, it should always be kept in mind that special parts of the structure that will be more exposed to intermittent wetting and drying than other parts of the structure will also be more vulnerable to corrosion of embedded steel.

5.3.7 Prefabricated structural elements

For many marine concrete structures, various types of prefabricated structural elements are being applied. Since the concrete construction work can then be carried out under more controlled conditions, this may also be a good strategy for protection against early-age chloride exposure during concrete construction. Possible surface protection systems to the most exposed concrete surfaces may further be applied under more controlled and optimal conditions before the prefabricated structural elements are installed.

For marine concrete construction, prefabricated structural elements may vary from small prefabricated formwork or deck slab elements to very large concrete structural elements. Some concrete structures may also partly or completely be prefabricated in a dry dock and later on moved into their final location. For many concrete bridges, large prefabricated deck beams and girders may be produced. Thus, for the Northumberland Strait Bridge Project on the east cost of Canada (Tromposch et al., 1998), both the pier shafts and the cantilever girders of up to 190 m length were prefabricated on shore before they were moved into position and installed by use of a heavy floating crane (Figures 5.5–5.8).

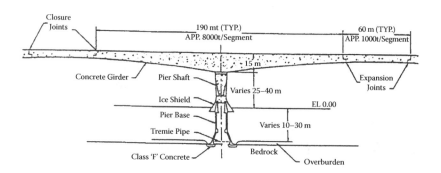

Figure 5.5 Typical foundation, pier shaft, and span for the Northumberland Strait Bridge (1997) on the east coast of Canada.

Figure 5.6 Precast cantilever girders for the Northumberland Strait Bridge.

Figure 5.7 Precast cantilever girder for the Northumberland Strait Bridge.

Figure 5.8 Installation of cantilever girders for the Northumberland Strait Bridge. (From Malhotra, V. M., *Proceedings, Third International Conference on Performance on Concrete in Marine Environment*, ACI SP-163, 1996.)

REFERENCES

ACI. (2007). *Report on Fiber-Reinforced Polymer (FRP) Reinforcement for Concrete Structures*, Report 440R-7. ACI Committee 440, Farmington Hills, MI.

ACMA MDA. (2006). http://www.mdacomposites.org.

Andrade, C. (1986). Some Laboratory Experiments on the Inhibitor Effect of Sodium Nitrite on Reinforcement Corrosion. *Cement and Concrete Aggregate*, 8(2), 110–116.

Arntsen, B. (2001). *In-Situ* Experiences on Chloride Redistribution in Surface-Treated Concrete Structures. In *Proceedings, Third International Conference on Concrete under Severe Conditions—Environment and Loading*, vol. 1, ed. N. Banthia, K. Sakai, and O.E. Gjørv. University of British Columbia, Vancouver, pp. 95–103.

Årskog, V., Borgund, K., and Gjørv, O.E. (2011). Effect of Concrete Hydrophobation against Chloride Penetration. *Key Engineering Materials*, 466, 183–190.

Årskog, V., Liu, G., Ferreira, M., and Gjørv, O.E. (2004). Effect of Surface Hydrophobation on Chloride Penetration into Concrete Harbor Structures. In *Proceedings, Fourth International Conference on Concrete under Severe Conditions—Environment and Loading*, vol. 1, ed. B.H. Oh, K. Sakai, O.E. Gjørv, and N. Banthia. Seoul National University and Korea Concrete Institute, Seoul, pp. 441–448.

Årstein, R., Rindarøy, O.E., Liodden, O., and Jenssen, B.W. (1998). Effect of Coatings on Chloride Penetration into Offshore Concrete Structures. In *Proceedings, Second International Conference on Concrete under Severe Conditions—Environment and Loading*, vol. 2, ed. O.E. Gjørv, K. Sakai, and N. Banthia. E & FN Spon, London, pp. 921–929.

Bertolini, L. (2000). Cathodic Prevention. In *Proceedings, COST 521 Workshop*, ed. D. Sloan and P.A.M. Basheer. Queen's University, Belfast.

Bertolini, L. (2012). Private communication.

Bertolini, L., Elsener, B., Pediferri, P., and Polder, R. (2004). *Corrosion of Steel in Concrete—Prevention, Diagnosis, Repair*. Wiley-VCH, Weinheim.

Bertolini, L., and Gastaldi, M. (2011). Corrosion Resistance of Low-Nickel Duplex Stainless Steel Rebars. *Materials and Corrosion*, 62, 120–129.

Bertolini, L., Lollini, F., Polder, R.B., and Peelen, W.H.A. (2008). FEM—Models of Chatodic Protection Systems for Concrete Structures. In *Proceedings, First International Symposium on Life-Cycle Civil Engineering*, ed. F. Biondini and D.M. Frangopol. Taylor & Francis Group, London, pp. 119–124.

Bertolini, L., Pediferri, P., and Pastore, T. (1998). Stainless Steel in Reinforced Concrete Structures. In *Proceedings, Second International Conference on Concrete under Severe Conditions—Environment and Loading*, vol. 1, ed. O.E. Gjørv, K. Sakai, and N. Banthia. E & FN Spon, London, pp. 94–103.

Bijen, J.M. (ed.). (1989). Maintenance and Repair of Concrete Structures. *Heron*, 34(2), 2–82.

Broomfield, J.P. (1997). *Corrosion of Steel in Concrete*. E & FN Spon, London.

Büchler, M. (2005). Corrosion Inhibitors for Reinforced Concrete. In *Corrosion in Reinforced Concrete Structures*, ed. H. Böhni. Woodhead Publishing, Cambridge, UK, pp. 190–214.

CEN. (1995). *EN 10088-1: Stainless Steels—Part 1: List of Stainless Steels*. European Standard CEN, Brussels.

CEN. (2000). *EN 12696: Cathodic Protection of Steel in Concrete*. European Standard CEN, Brussels.

CSA. (2003). *S806-02: Fibre Reinforced Polymer Reinforcement for Concrete Structures*. Canadian Standards Association, Rexdale, Ontario.

Clemeña, G.G. (2002). *Testing of Selected Metallic Reinforcing Bars for Extending the Service Life of Future Concrete Bridges: Summary of Conclusions and Recommendations*, Report VTRC 03-R7. Virginia Transportation Research Council, Charlottesville.

Clemeña, G.G., and Virmani, Y.P. (2004). Comparing the Chloride Resistances of Reinforcing Bars. *Concrete International*, 26(11), 39–49.

The Concrete Society. (1998). *Guidance on the Use of Stainless Steel Reinforcement*, Technical Report 51. London.

COST. (2003). *COST Action 521: Corrosion of Steel in Reinforced Concrete Structures*, Final Report, ed. R. Cigna, C. Andrade, U, Nürnberger, R. Polder, R. Weydert. and E. Seitz. European Communities EUR20599, Luxemborg.

Cramer, S.D., Covino, B.S., Bullard, S.J., Holcomb, G.R., Russell, J.H., Nelson, F.J., Laylor, H.M., and Stoltesz, S.M. (2002). Corrosion Prevention and Remediation Strategies for Reinforced Concrete Coastal Bridges. *Cement and Concrete Composites*, 24, 101–117.

De Clereq, H., and Charola, A.E. (2008). *Proceedings, Fifth International Conference on Water Repellent Treatment of Building Materials*. Aedificatio Publishers, Freiburg.

FIB. (2007). *FRP Reinforcement in RC Structures*, FIB Bulletin 40. Lausanne.

FIP. (1996). *Durability of Concrete Structures in the North Sea*, State-of-the-Art Report. Féderation Internationale de la Précontrainte—FIP, London.

Gaidis, J.M., and Rosenberg, A.M. (1987). The Inhibition of Chloride-Induced Corrosion in Reinforced Concrete by Use of Calcium Nitrite. *Cement and Concrete Aggregate*, 9(1), 30–33.

Gaubinger, B., Bahr, G., Hampel, G., and Kollegger, J. (2002). Innovative Anchorage System for GFRP-Tendons. In *Proceedings, First FIB Congress*, session 7, vol. 6. Japan Prestressed Concrete Engineering Association, Tokyo, pp. 305–312.

Gjørv, O.E., and Vennesland, Ø. (1987). Evaluation and Control of Steel Corrosion in Offshore Concrete Structures. In *Proceedings, The Katharine and Bryant Mather International Conference*, vol. 2, ed. J. Scanlon, ACI SP-1, pp. 1575–1602.

Hinatsu, J.T., Graydon, W.F., and Foulkes, F.R. (1990). Voltametric Behaviour of Iron in Cement. Effect of Sodium Chloride and Corrosion Inhibitor Additions. *Journal of Applied Electrochemistry*, 20(5), 841–847.

Isaksen, T. (2004). The Owners Profit by Investing in Increased Durability. In *Proceedings, Seminar on Service Life Design of Concrete Structures*. Norwegian Concrete Association, TEKNA, Oslo, pp. 15.1–15.10 (in Norwegian).

Knudsen, A., and Goltermann, P. (2004). Stainless Steel as Reinforcement for Concrete. In *Proceedings, Seminar on Service Life Design of Concrete Structures*. Norwegian Concrete Association, TEKNA, Oslo, pp. 10.1–10.12 (in Danish).

Knudsen, A., Jensen, F.M., Klinghoffer, O., and Skovsgaard, T. (1998). Cost-Effective Enhancement of Durability of Concrete Structures by Intelligent Use of Stainless Steel Reinforcement. In *Proceedings, Conference on Corrosion and Rehabilitation of Reinforced Concrete Structures*, Florida.

Liu, G. (2006). Control of Chloride Penetration into Concrete Structures at Early Age, Dr.Ing. Thesis 2006:46. Department of Structural Engineering, Norwegian University of Science and Technology—NTNU, Trondheim.

Liu, G., Stavem, P., and Gjørv, O.E. (2005). Effect of Surface Hydrophobation for Protection of Early Age Concrete against Chloride Penetration. In *Proceedings, Fourth International Conference on Water Repellent Treatment of Building Materials*, ed. J. Silfwerbrand. Aedificatio Publishers, Freiburg, pp. 93–104.

LME—London Metal Exchange. www.lme.com.

Malhotra, V.M. (ed.). (1996). *Proceedings, Third International Conference on Performance on Concrete in Marine Environment*, ACI SP-163.

Materen, S. von, and Paulsson-Tralla, J. (2001). The De-Icing Salt—Stop the Damage on Exposed Concrete Structures by Use of Stainless Steel Bars. *Betong*, 2, 18–22 (in Swedish).

Newhook, J. (2006). Glass FRP Reinforcement in Rehabilitation of Concrete Marine Infrastructure. *Arabian Journal for Science and Engineering*, 31(1C), 53–75.

Newhook, J., Bakht, B., Tadros, G., and Mufti, A. (2000). Design and Construction of a Concrete Marine Structure Using Innovative Technologies. In *Proceedings of the Third International Conference on Advanced Composite Materials in Bridges and Structures (CSCE)*, Ottawa, ed. M. El-Badry.

Newhook, J., and Mufti, A. (1996). A Reinforcing Steel-Free Concrete Bridge Deck for the Salmon River Bridge. *Concrete International*, 18.

Ngala, V.T., Page, C.L., and Page, M.M. (2002). Corrosion Inhibitor Systems for Remedial Treatment of Reinforced Concrete. I. Calcium Nitrite. *Corrosion Science*, 44(9), 2073–2087.

Noisternig, F., Dotzler, F., and Jungwirth, D. (1998). *Development of CFK Prestressed Elements*. Seminarband Kreative Ingenieurleistungen, Darmstadt-Wien (in German).

Nürnberger, U. (1988). *Special Measures for Corrosion Protection of Reinforced and Prestressed Concrete*, Otto-Graf-Institute Series 79. Stuttgart (in German).

Pediferri, P. (1992). Cathodic Protection of New Concrete Construction. In *Proceedings, International Conference on Structural Improvement through Corrosion Protection of Reinforced Concrete*, Document E7190. Institute of Corrosion, London.

Pediferri, P. (2006). *Surface Rebar Contamination*, Report. Acciaierie Valbruna, Vicenza, Italy.

Pediferri, P., Bertolini, L., Bolzoni, F., and Pastore, T. (1998). In *Repair and Rehabilitation of Reinforced Concrete Structures: State of the Art*, ed. W.F. Silva Araya, O.T. De Rincon, and L.P. O'Neill. American Society of Civil Engineering, Reston, VA.

Polder, R.B., Leegwater, G., Worm, D., and Courage, W. (2012). *Service Life and Life Cycle Cost Modelling of Cathodic Protection Systems for Concrete Structures*. International Congress on Durability of Concrete, Norwegian Concrete Association, Oslo.

Rasheeduzzafar, F.H.D., Bader, M.A., and Kahn, M.M. (1992). Performance of Corrosion Resisting Steels in Chloride-Bearing Concrete. *ACI Materials Journal*, 89(5), 439–448.

Raupach, M., and Rößler, G. (2005). Surface Treatments and Coatings for Corrosion Protection. In *Corrosion in Reinforced Concrete Structures*. Woodhead Woodhead Publishing, Cambridge, UK, pp. 163–189.

ReforceTech. (2013). Basalt Fiber Reinforcement Technology. http://www.reforcetech.com.

Serancio, R. (ed.). (2004). FRP in Civil Engineering. In *Proceedings of the Second International Conference on FRP Composites in Civil Engineering*. Balkema Publishers, London.

Silfwerbrand, J. (ed.). (2005). *Proceedings, Fourth International Conference on Water Repellent Treatment of Building Materials*. Aedificatio Publishers, Freiburg, Germay.

Sørensen, B., Jensen, P.B., and Maahn, E. (1990). The Corrosion Properties of Stainless-Steel Reinforcement. In *Corrosion of Reinforcement in Concrete*, ed. C.L. Page, K.W.J. Treadaway, and P.B. Bamforth. Elsevier Applied Science, Amsterdam, pp. 601–610.

Tan, K. (ed.). (2003). *Proceedings of the Sixth International Symposium of FRP Reinforcement for Concrete Structures*. World Scientific, Singapore.

Tromposch, E.W., Dunaszegi, L., Gjørv, O.E., and Langley, W.S. (1998). Northumberland Strait Bridge Project—Strategy for Corrosion Protection. In *Proceedings, Second International Conference on Concrete under Severe Conditions—Environment and Loading*, vol. 3, ed. O.E. Gjørv, K. Sakai, and N. Banthia. E & FN Spon, London, pp. 1714–1720.

Chapter 6

Concrete quality control and quality assurance

6.1 GENERAL

In order to obtain better control of the high scatter and variability of achieved construction quality typically observed (Chapter 2), it is essential to have some performance-based durability requirements that can be verified and controlled for quality assurance during concrete construction. Even for the offshore concrete structures where the strictest procedures for both concrete production and quality control during concrete construction were applied, these concrete structures also partly showed a high scatter and variability of achieved construction quality. Although the regular concrete quality control typically revealed a very homogeneous concrete production (Table 2.1), a high scatter and variability of achieved in situ quality of the concrete could still be observed (Figure 2.61).

Even before the concrete is placed in the formwork, the concrete quality may show a high scatter and variability (Gulikers, 2011). Depending on a number of factors during concrete construction, the achieved quality of the finely placed concrete may show an even higher scatter and variability. If air-entrained concrete is applied, large variations in the air-void characteristics may also occur (Gjørv and Bathen, 1987). This problem is exacerbated when fly ash cements are used and the carbon content of the fly ash varies (Nagi et al., 2007), affecting both the chloride diffusivity and the frost resistance of the concrete.

Probably the best-known and well-documented quality issue of concrete structures is the failure to meet the specified requirements to concrete cover. The variations of achieved concrete cover in the Norwegian bridge shown in Figure 2.40 reflect a general problem on many construction sites in many countries. Thus, in Figure 6.1, the data from this bridge are plotted together with similar data from a Japanese bridge (Ohta et al., 1992) and the average of data from more than 100 concrete structures in the Gulf region (Matta, 1993). In recent years, therefore, improved codes and procedures for achieving the specified concrete cover with more confidence have been introduced. Still, however, the variability of concrete

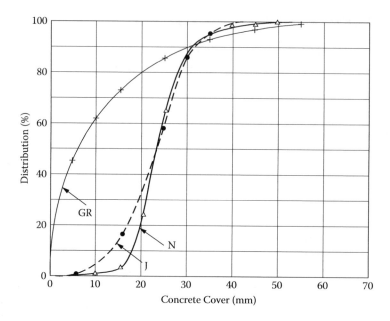

Figure 6.1 Variation in achieved concrete cover (mm) in the Gimsøystraumen Bridge (N) previously shown in Figure 2.40 compared to that of a Japanese bridge (J) and more than 100 concrete structures in the Gulf region (GR). (From Kompen, R., What Can Be Done to Improve the Quality of New Concrete Structures? in *Proceedings, Second International Conference on Concrete under Severe Conditions—Environment and Loading*, vol. 3, ed. O. E. Gjørv et al., E & FN Spon, London, England, 1998, pp. 1519–1528.)

cover appears to be a very difficult problem. Although the specified concrete cover is normally carefully checked and controlled before placing of the concrete, experience demonstrates that significant deviations can still occur during concrete construction, and this does occur. The loads during placing of the concrete may occasionally be too high compared to the stiffness of the rebar system, or the spacers may occasionally have been insufficiently or wrongly placed. Even during the sophisticated slip forming work for the offshore concrete platforms, the installed spacers occasionally were removed during some critical stages of the slip forming in order to keep the slip forming work going on.

In order to comply with the overall durability requirement to the given concrete structure as described in Chapter 4, the specified values of both the 28-day chloride diffusivity and the concrete cover must be properly controlled, which is achieved through ongoing verification and documentation during concrete construction. For both of these durability parameters, average values and standard deviations must be obtained, and when unacceptable deviation occurs, immediate actions for correction must be taken.

6.2 CHLORIDE DIFFUSIVITY

6.2.1 General

As already outlined and discussed in Chapter 4, all testing of chloride diffusivity is based on the rapid chloride migration (RCM) method (NORDTEST, 1999). Although the duration of such a test may only take a few days, this is not good enough for regular quality control and quality assurance during concrete construction. For any porous materials, however, the Nernst-Einstein equation expresses the following general relationship between the diffusivity and the electrical resistivity of the material (Atkins and De Paula, 2006):

$$D_i = \frac{R \cdot T}{Z^2 \cdot F^2} \cdot \frac{t_i}{\gamma_i \cdot c_i \cdot \rho} \qquad (6.1)$$

where D_i = diffusivity for ion i, R = gas constant, T = absolute temperature, Z = ionic valence, F = Faraday constant, t_i = transfer number of ion i, γ_i = activity coefficient for ion i, c_i = concentration of ion i in the pore water, and ρ = electrical resistivity.

Since most of the factors in Equation 6.1 are physical constants, the above relationship can, for a given concrete with given temperature and moisture conditions, be simplified to

$$D = k \cdot \frac{1}{\rho} \qquad (6.2)$$

where D is the chloride diffusivity, k is a constant, and ρ is the electrical resistivity of the concrete. Since the electrical resistivity of the concrete can be tested in a very rapid and simple way compared to that of the chloride diffusivity, it is primarily a regular quality control of the electrical resistivity of the concrete, which provides the basis for an indirect control and quality assurance of the chloride diffusivity during concrete construction (Gjørv, 2003). Therefore, the above relationship between chloride diffusivity and electrical resistivity for the given concrete must be established before the concrete construction starts. This is done by producing a certain number of concrete specimens, from which parallel testing of chloride diffusivity and electrical resistivity are carried out at different periods of water curing at 20°C, an example of which is shown in Figure 6.2.

After the above relationship between the chloride diffusivity and the electrical resistivity of the given concrete has been established, it is later used as a calibration curve for indirect control of the 28-day chloride diffusivity based on regular testing of the electrical resistivity during concrete construction. Since this testing of electrical resistivity is a nondestructive type

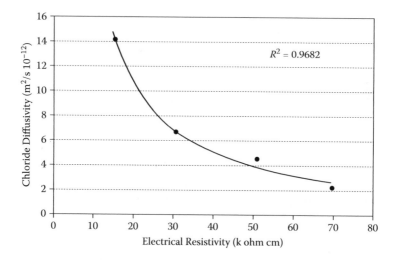

Figure 6.2 A typical calibration curve for control of chloride diffusivity based on measurements of electrical resistivity.

of test, these measurements are carried out as a very quick test on the same concrete specimens as those being used for the regular quality control of the 28-day compressive strength. Of all these control data, any individual value should never exceed 30% by that of the specified chloride diffusivity.

In order to establish the above calibration curve, the measurements of chloride diffusivity are carried out on three 50 mm thick specimens cut from Ø100 × 200 mm concrete cylinders after approximately 7, 14, 28, and 60 days of water curing. After cutting, the specimens are water saturated according to the established water saturation procedure before further testing. In parallel, the corresponding measurements of the electrical resistivity are carried out on three concrete specimens of the same type as that being used for the regular quality control of compressive strength. These measurements are carried out on moist concrete specimens after the above curing periods in water.

It is not the purpose here to describe all the details for the measurements of chloride diffusivity based on the RCM method. Such measurements require special testing equipment and qualified experience, which are only available in professional testing laboratories. However, for a better evaluation and application of the obtained results, a brief outline of the test method is given below.

6.2.2 Test specimens

The testing is normally based on 3 × 50 mm cut slices of a concrete cylinder with diameter Ø100 mm. The 50 mm thick slices are cut either from separately cast concrete cylinders or from concrete cores removed from either

the structure under construction or corresponding dummy elements produced on the construction site.

6.2.3 Testing procedure

Immediately before testing, the test specimens are preconditioned according to a standardized water saturation procedure. Then, the specimens are
mounted into rubber sleeves and placed in a container with a 10% NaCl
solution, as shown in Figure 6.3, while the inside of the sleeves are filled
with a 0.3 N NaOH solution. A set of concrete specimens under testing
with the RCM method is shown in Figure 6.4.

By use of the separate electrodes placed on each side of the specimens, an
electrical voltage gradient is applied and the electrical current through the
specimens observed. Depending on the level of the observed current that
reflects the resistance of the concrete to chloride ingress, the applied potential is adjusted in order to obtain a proper duration of the testing. By an
applied DC potential that may vary from 10 to 60 V, the chloride ions are
forced into the concrete specimens during a relatively short period of time.
For a normal dense concrete, a testing duration of 24 hours may be typical,
while for a denser concrete, a longer time is needed.

Immediately after termination of the accelerated exposure to the chloride
solution, the test specimens are split into two halves, from which the average depth of chloride penetration is observed on the freshly split surface by

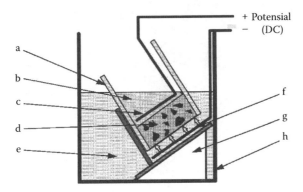

Figure 6.3 Experimental setup for the RCM testing of chloride diffusivity (D_0): (a) rubber sleeve, (b) anolyte, (c) anode, (d) concrete specimen, (e) catholyte,
(f) cathode, (g) plastic support, and (h) plastic box. (From NORDTEST, *NT
Build 492 Concrete, Mortar, and Cement Based Repair Materials: Chloride Migration
Coefficient from Non-Steady State Migration Experiments*, NORDTEST, Espoo,
Finland, 1999; Tang, *The RCM Test (NT Build 492) for Evaluating the Resistance
of Concrete to Chloride Ingress*, Tang's CI Tech, Gothenburg, Sweden, 2008,
tang.luping@bredband.net.)

Figure 6.4 Concrete specimens under testing with the RCM method. (From NORDTEST, *NT Build 492 Concrete, Mortar, and Cement Based Repair Materials: Chloride Migration Coefficient from Non-Steady State Migration Experiments*, NORDTEST, Espoo, Finland, 1999; Tang, *The RCM Test (NT Build 492) for Evaluating the Resistance of Concrete to Chloride Ingress*, Tang's CI Tech, Gothenburg, Sweden, 2008, tang.luping@bredband.net.)

Figure 6.5 Observed depth of chloride penetration on the split concrete surface after spraying of the fresh surface with a standard $AgNO_3$ solution.

use of a colorimetric technique (Figure 6.5). Based on the observed depth of chloride ingress and the applied testing conditions, the chloride diffusivity is calculated according to an established procedure. As a result, the chloride diffusivity (D) of the concrete is obtained as an average value and standard deviation from the testing of three test specimens. Depending on the resistance to chloride ingress of the given concrete, the whole test period, including preparation of the concrete specimens, may take a few days.

6.2.4 Evaluation of obtained results

When the resistance of the concrete to chloride ingress is determined and characterized by such a strongly accelerated test method as that described above, it should be clear that the results obtained are quite different from what takes place under more normal conditions in the field. Therefore, from both a transport mechanism and a theoretical point of view, it may be argued that an input parameter based on such a test method is questionable (Gulikers, 2011; Yuan, 2009). It should be noted, however, that the above 28-day chloride diffusivity is only a simple, relative index reflecting the general mobility of ions in the pore system of the concrete, and hence both the resistance to chloride ingress and the general durability properties of the concrete. Thus, the 28-day chloride diffusivity may be comparable to that of the 28-day compressive strength, which is also only a very simple, relative index reflecting the compressive strength as well as the general mechanical properties of the concrete.

Since RCM is the only method that requires a short duration of testing, it is the only test method available for control of chloride diffusivity at an early stage of hydration, independent of concrete age. For a proper durability design of concrete structures in severe environments, the 28-day chloride diffusivity is a very important parameter. The specification of a given 28-day chloride diffusivity also provides the necessary basis for the regular concrete quality control and quality assurance during concrete construction.

For a more complete evaluation and comparison of the resistance to chloride ingress of various types of concrete, however, not only the 28-day chloride diffusivity should be evaluated. It is the more complete development of chloride diffusivity from an early age and up that more properly reflects the durability properties of the given concrete, and hence the different resistance to chloride ingress of various types of concrete.

6.3 ELECTRICAL RESISTIVITY

6.3.1 General

Also for the testing of the electrical resistivity of concrete, several experimental techniques and test methods exist, all of which give different test results (Gjørv et al., 1977; Polder et al., 2000). Basically, however, there are two different types of test methods that appear suitable for regular quality control during concrete construction, one of which is the two-electrode method, and the other the four-electrode method, or the so-called Wenner method.

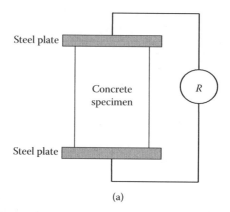

(a)

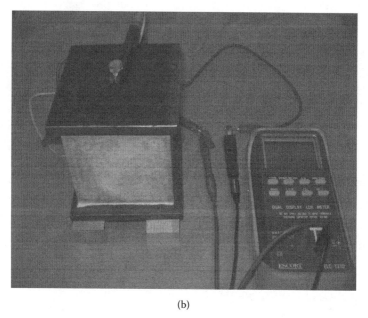

(b)

Figure 6.6 The two-electrode method for electrical resistivity measurements of concrete.

6.3.2 Test methods

Schematically, the two-electrode method is shown in Figure 6.6, where the applied current flows through the whole bulk of the concrete specimen, which is placed between two solid steel plates; the resistance is then observed by use of a resistance meter at a certain frequency (1 kHz). The resistivity of the concrete (ρ) is calculated by the following equation:

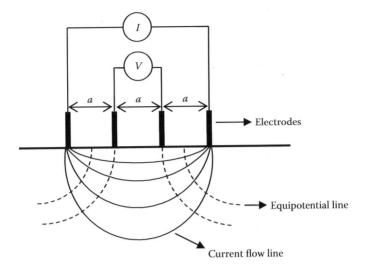

Figure 6.7 The four-electrode method (Wenner method) for electrical resistivity measurements of concrete.

$$\rho = R \frac{A}{t} \tag{6.3}$$

where R is the resistance, A is the surface area of the specimen, and t is the height of the concrete specimen.

The four-electrode method (Wenner) is schematically shown in Figure 6.7. For this method, the measurements are based on a low-frequency alternating electrical current passing through the concrete between the two outer electrodes, while the voltage drop between the two inner electrodes is observed. The electrical resistivity of the concrete (ρ) is then obtained by the following equation:

$$\rho = 2\pi a V/I \tag{6.4}$$

where a is the electrode spacing, V is the voltage drop, and I is the current.

As shown in Figures 6.8 and 6.9, the Wenner method consists of a device with four electrodes that is pressed against the concrete surface, and the apparent resistivity is then observed on a display for the given electrode distance (*a*). For some commercial units, the electrode distance is adjustable.

For the Wenner method, it appears from Figure 6.7 that the flow of current through the concrete is very much affected by both the electrode spacing and the geometry of the concrete specimen. For this test method,

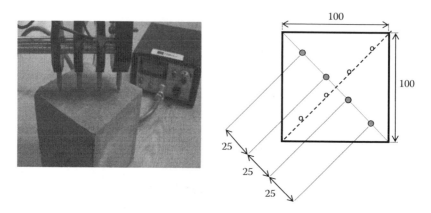

Figure 6.8 Measurement of electrical resistivity on a 100 mm concrete cube (mm).

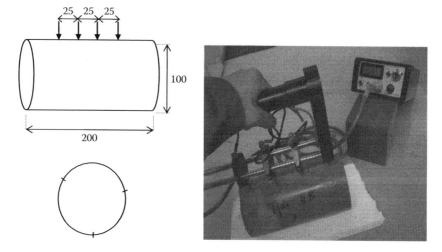

Figure 6.9 Measurement of electrical resistivity on Ø100 × 200 mm concrete cylinders (mm).

therefore, it is very important to keep the same testing procedure constant throughout the quality control.

6.3.3 Evaluation of obtained results

Since the two-electrode method is the most well-defined and accurate test method, this method should preferably be applied for the regular concrete quality control; the two-electrode method is also less dependent on the operator. However, both the above methods can be applied for a simple and rapid quality control during concrete construction. As can be seen from Figure 6.10, the correlation between the results obtained by the two test

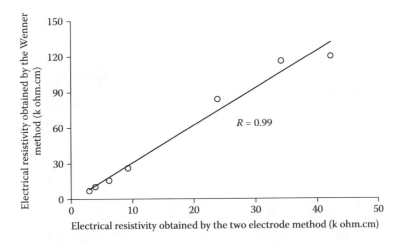

Figure 6.10 Relationship between the resistivities obtained by the Wenner method and the two-electrode method. (From Sengul, Ő., and Gjørv, O. E., Electrical Resistivity Measurements for Quality Control during Concrete Construction, *ACI Materials Journal*, 105, 541–547, 2008.)

methods is also good. As soon as the calibration curve is established based on any of the above test methods, however, it is essential to keep the same test method and testing procedure constant throughout the further quality control.

Since both the moisture and temperature conditions are very important factors affecting the electrical resistivity of the concrete, all measurements of the electrical resistivity must be carried out under controlled conditions in the laboratory. Immediately before testing, all free water on the surface of the concrete specimen must be carefully wiped off. Also, it is very important to ensure a good electrical connection between the electrodes and the concrete surface. In order to avoid any current drain during measurements, the specimens must further be placed on a dry, electrically insulated base plate, and any touching of the concrete specimen by hand avoided.

6.4 CONCRETE COVER

For concrete structures in severe environments, the specified concrete cover is normally very thick, and the reinforcement is often highly congested, making it difficult to measure the cover thickness accurately based on conventional cover meters. The use of stainless steel reinforcement may further complicate such measurements, although cover meters based on pulse induction can then be used (Figure 6.11). Also, more sophisticated scanning systems for control of achieved concrete cover exist (Figure 6.12). With

Figure 6.11 Control of achieved concrete cover based on the pulse induction technique. (From Germann, *In-Situ Test Systems for Durability, Inspection, and Repair of Reinforced Concrete Structures*, Germann Instruments A/S, Copenhagen, Denmark, 2008, http://www.germann.org/.)

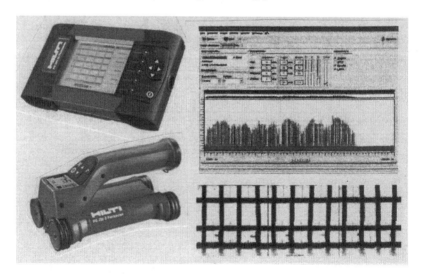

Figure 6.12 Control of achieved concrete cover based on scanning systems. (From Grantham, M., The Use of NDT Methods in the Evaluation of Structures, in *Proceedings, Second International Symposium on Service Life Design for Infrastructure*, vol. 2, RILEM, Bagneux, France, 2010, pp. 1003–1012.)

such equipment available, there should be little excuse for allowing low cover in new concrete structures, yet this seems to continue to be a problem on many construction sites.

Very often, a more pragmatic approach based on manual readings of the cover thickness on protruding bars in construction joints during concrete construction is applied. If the amount of such control measurements is sufficiently high to produce reliable statistical data, such a simple approach may be sufficiently accurate for the regular control and quality assurance during concrete construction. As long as ongoing control and documentation of the achieved concrete cover are required, experience has shown that the increased focus and attention of achieving the specified cover in itself are also very important for an increased quality of workmanship (Chapter 9).

6.5 ELECTRICAL CONTINUITY

6.5.1 General

If cathodic prevention or provisions for such a protective measure should be specified, requirements to the electrical continuity within the reinforcement system are given in the European standard for cathodic protection of concrete structures, EN 12696 (CEN, 2000). For heavily reinforced concrete structures, there may already be a sufficient electrical continuity without any additional measures, but normally, both welding and special electrical connections between various parts of the rebar system are needed. For each step of the concrete construction work, however, the specified electrical continuity of the reinforcement must be properly controlled and secured. Such control should preferably be carried out and assured by qualified people with experience with cathodic protection systems.

6.5.2 Testing procedure

In principle, the electrical continuity is controlled by measurements of the ohmic resistance between various parts of the reinforcement system, and between any two points in the system, the resistance shall not exceed 1 ohm. In order to avoid the uncertainty of measurements when a traditional multimeter is applied, the measurements should preferably be carried out by use of a relatively high current (1 A) between the various parts of the reinforcement system. Commercial equipment for such measurements is available, where both the ohmic resistance and the rest voltage between the measuring parts 0.1 second after interruption of the current are observed (Figure 6.13).

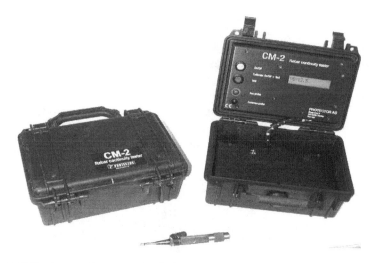

Figure 6.13 Equipment for control of electrical continuity within the reinforcement system. (From Protector, *CM2 Rebar Continuity Tester*, Protector AS, Oslo, Norway, 2008, http:// www.protector.no/.)

REFERENCES

Atkins, P.W., and De Paula, J. (2006). *Physical Chemistry*, 8th ed. Oxford University Press, Oxford.

CEN. (2000). *EN 12696: Cathodic Protection of Steel in Concrete*. European Standard, CEN, Brussels.

Germann. (2008). *In-Situ Test Systems for Durability, Inspection and Repair of Reinforced Concrete Structures*. Germann Instruments A/S, Copenhagen, www.germann.org/.

Gjørv, O.E. (2003). Durability of Concrete Structures and Performance-Based Quality Control. In *Proceedings, International Conference on Performance of Construction Materials in the New Millennium*, ed. A.S. El-Dieb, M.M.R. Taha, and S.L. Lissel. Shams University, Cairo.

Gjørv, O.E., and Bathen, E. (1987). Quality Control of the Air-Void System in Hardened Concrete. *Nordic Concrete Research*, 6, 95–110.

Gjørv, O.E., Vennesland, Ø., and El-Busaidy, A.H.S. (1977). Electrical Resistivity of Concrete in the Oceans, OTC Paper 2803. In *Annual Offshore Technology Conference*, Houston, TX, pp. 581–588.

Grantham, M. (2010). The Use of NDT Methods in the Evaluation of Structures. In *Proceedings, Second International Symposium on Service Life Design for Infrastructure*, vol. 2. RILEM, Bagneux, France, pp. 1003–1012.

Gulikers, J. (2011). Practical Implications of Performance Specifications for Durability Design of Reinforced Concrete Structures. In *Proceedings, International Workshop on Performance-Based Specifications for Concrete*, ed. F. Dehn and H. Beushausen. MFPA Leipzig GmbH, Institute for Materials Research and Testing, Leipzig, pp. 341–350.

Kompen, R. (1998). What Can Be Done to Improve the Quality of New Concrete Structures? In *Proceedings, Second International Conference on Concrete under Severe Conditions—Environment and Loading*, vol. 3, ed. O.E. Gjørv, K. Sakai, and N. Banthia. E & FN Spon, London, pp. 1519–1528.

Matta, Z.G. (1993). Deterioration of Concrete Structures in the Arabian Gulf. *Concrete International*, 15, 33–36.

Nagi, M.A., Okamoto, P.A., Kozikowski, R.L., and Hover, K. (2007). *Evaluating Air-Entraining Admixtures for Highway Concrete*, NCHRP Report 578. Transportation Research Board, Washington, D.C.

NORDTEST. (1999). *NT Build 492 Concrete, Mortar and Cement Based Repair Materials: Chloride Migration Coefficient from Non-Steady State Migration Experiments*. NORDTEST, Espoo, Finland.

Ohta, T., Sakai, K., Obi, M., and Ono, S. (1992). Deterioration in a Rehabilitated Prestressed Concrete Bridge. *ACI Materials Journal*, 89, 328–336.

Polder, R., Andrade, C., Elsener, B., Vennesland, Ø., Gulikers, J., Weidert, R., and Raupach, M. (2000). RILEM TC 154-EMC: Electrochemical Techniques for Measuring Metallic Corrosion. *Materials and Structures*, 33, 603–611.

Protector. (2008). *CM2 Rebar Continuity Tester*. Protector AS, Oslo, www.protector. no/.

Sengul, Ő., and Gjørv, O.E. (2008). Electrical Resistivity Measurements for Quality Control during Concrete Construction. *ACI Materials Journal*, 105, 541–547.

Tang. (2008). *The RCM Test (NT Build 492) for Evaluating the Resistance of Concrete to Chloride Ingress*. Tang's Cl Tech, Gothenburg, tang.luping@bredband.net.

Yuan, Q. (2009). Fundamental Studies on Test Methods for the Transport of Chloride Ions in Cementitious Materials, PhD Thesis. University of Ghent, Ghent.

Chapter 7

Achieved construction quality

7.1 GENERAL

As a result of the regular concrete quality control (Chapter 6), average values and standard deviations of both the 28-day chloride diffusivity of the concrete and the concrete cover are obtained. Upon completion of the concrete construction, therefore, these data are used as input parameters to a new durability analysis in order to provide documentation of compliance with the specified durability.

Since the specified chloride diffusivity is only based on small and separately produced concrete specimens cured in the laboratory for 28 days, this chloride diffusivity may be quite different from that obtained on the construction site during the construction period. Therefore, additional documentation on the achieved chloride diffusivity on the construction site during the construction period must also be obtained. At the end of concrete construction, this chloride diffusivity in combination with the achieved concrete cover provides the basis for documentation of the achieved in situ quality.

Since neither the 28-day chloride diffusivity from the laboratory nor the achieved in situ diffusivity on the construction site reflects the potential chloride diffusivity of the given concrete, further documentation on the long-term diffusivity of the given concrete must also be obtained. This chloride diffusivity in combination with the achieved concrete cover provides the basis for documentation of the potential construction quality of the given concrete structure.

Based on the above durability analyses, it should be noted that the achieved construction quality is being expressed and quantified as a probability of corrosion for the required service period.

For the owner of the structure, proper documentation of the achieved construction quality and compliance with the specified durability should be very important, since this may have implications for both the future operation and expected service life of the given structure. In the following, the procedures for providing such documentation are described in more detail.

7.2 COMPLIANCE WITH SPECIFIED DURABILITY

As an overall durability requirement to the given concrete structure, a required service period with a corrosion probability of less than 10% is specified. To show compliance with such a specification, a new durability analysis must be carried out upon completion of the concrete construction period, and this analysis must be based on the achieved average values and standard deviations of both the 28-day chloride diffusivity and the concrete cover as new input parameters. For this durability analysis, all the other previously assumed input parameters, which may have been somewhat difficult to select during durability design, are now kept constant. Therefore, this documentation primarily reflects the results obtained from the regular quality control of 28-day chloride diffusivity and concrete cover during concrete construction, including the observed scatter and variability. Hence, this new durability analysis provides the basis for documentation of compliance with the specified durability.

7.3 IN SITU QUALITY

In order to obtain data on the achieved chloride diffusivity on the construction site during concrete construction, a number of concrete cores have to be removed from the given structure under construction. In order not to weaken the structure too much by coring, however, one or more representative dummy elements should also be produced on the construction site, from which a number of additional cores are removed and tested; these dummy elements are normally produced without any reinforcement. Although it may not be easy to reflect the same conditions as in the real structure, it is essential that all the separately produced dummy elements, which can be either a wall or a slab type of element, or both, are produced and cured as representative as possible for the real concrete structure or various parts of the concrete structure under construction.

From both the concrete structure and the corresponding dummy elements, a number of Ø100 mm concrete cores are removed at various stages during the construction period and, immediately upon removal, sent to the laboratory for the testing of chloride diffusivity. Somewhat depending on the type of binder system and the given curing conditions at the construction site, the development of this in situ chloride diffusivity often tends to level out after a period of approximately one year. In order to obtain a proper curve for the development of chloride diffusivity, the cores should be removed and tested during a period of up to at least one year, preferably after periods of approximately 14, 28, 60, 90, 180, and 365 days. As an example, the observed development of achieved chloride diffusivity from one particular construction site is shown in Figure 7.1. In the same figure,

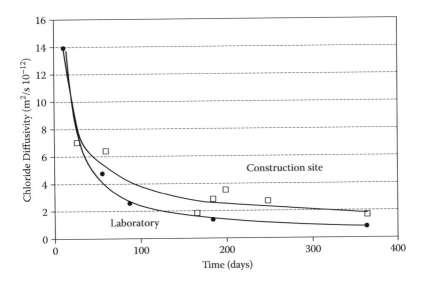

Figure 7.1 Developments of achieved chloride diffusivity on the construction site and in the laboratory during the construction period.

the development of chloride diffusivity for the same concrete based on separately cast and cured concrete specimens in the laboratory is also shown.

Based on the achieved chloride diffusivity on the construction site after one year, combined with the achieved site data on cover thickness as input parameters, a new durability analysis is carried out. Also here, all the other previously assumed input parameters are kept constant. Hence, this analysis provides the basis for documentation of the achieved in situ quality during the construction period.

7.4 POTENTIAL QUALITY

Already during the establishment of the calibration curve for regular concrete quality control, the chloride diffusivity is determined on separately cast concrete specimens after periods of water curing in the laboratory of approximately 7, 14, 28, and 60 days. By continued testing of the chloride diffusivity on a few additional concrete specimens after further curing periods in the laboratory of approximately 90, 180, and up to at least 365 days, a further development of the chloride diffusivity, like that shown in Figure 7.1, is typically obtained.

Although it may take a long time before a final value of the chloride diffusivity is reached, this development curve will, for most binder systems, tend to plateau after about one year of water curing. Based on the achieved chloride diffusivity after one year of water curing in the laboratory, combined with the achieved site data on cover thickness as new input parameters, another durability analysis is carried out. Also here, all the other previously assumed input parameters are kept constant. Hence, this analysis provides the basis for documentation of the potential construction quality of the given structure.

Chapter 8

Condition assessment, preventive maintenance, and repairs

8.1 GENERAL

For the operation of most existing concrete structures, the typical situation is that maintenance and repairs are mostly reactive, and the need for taking appropriate measures is mostly realized at a very advanced stage of deterioration (Chapter 2). For chloride-induced corrosion, repairs at such a stage are then both technically difficult and disproportionately costly compared to that of carrying out regular condition assessments and preventive maintenance during operation of the structures. Therefore, for all major concrete infrastructures where high safety, performance, and service life are of special importance, regular condition assessment and preventive maintenance should be carried out. Not only is this important from a technical and economical point of view, but also this has shown to be a very good strategy from a sustainability point of view (Chapter 11).

In recent years, a rapid international development on general systems for life cycle management (LCM) of important infrastructure facilities has taken place (Grigg, 1988; O'Connor and Hyman, 1989; Hudson et al., 1987; RIMES, 1997; BRIME, 1997). Depending on the number of facilities to be included, established LCM systems for both network and project levels are commercially available. In many countries, national authorities have also developed their own LCM systems. For many important concrete structures, therefore, general condition assessment and preventive maintenance are already part of the established LCM systems.

For all important concrete structures in chloride-containing environments, however, special procedures for control of chloride ingress are needed, and the establishment of such procedures should always be an important and integral part of the durability design for the given structure (Eri et al., 1998; Tromposch et al., 1998; Gjørv, 2002). Already at an early stage, appropriate locations for future control of chloride ingress in critical parts of the given structure should be selected, and such locations should be easily accessible during operation of the structure. If critical parts of the structure should not be accessible for future testing of chloride ingress, however, instrumentation based on embedded probes for chloride monitoring may be considered. If cathodic prevention or provisions for such a protective measure should be applied, necessary specifications for this must also be established at an early stage of the durability design. A regular control of chloride ingress during operation of the given structure is very important, the general basis for which is briefly described and outlined below. If a stage of corrosion should be reached where repairs are needed, a brief outline of current experience with such repairs is also included.

8.2 CONTROL OF CHLORIDE INGRESS

Even if the strongest requirements to both concrete quality and concrete cover have been specified and achieved during concrete construction, extensive experience demonstrates that for concrete structures in marine environments, a varying extent of chloride ingress will always take place during operation of the structures. For concrete construction work in marine environments, an early-age chloride exposure may also take place before the concrete has gained sufficient curing and maturity, as shown in Chapter 2. If the risk for such early-age exposure during concrete construction is high, this must also be properly considered and taken into account during durability design. In such a case, an early control of possible chloride ingress during concrete construction may also be required before the concrete structure is handed over from the contractor.

For the regular condition assessment and control of chloride ingress during operation of the structure, it is very important to have a detailed plan for the given structure showing the selected locations where the future control shall take place. These locations must be as representative as possible for the most exposed and critical parts of the structure. Since all concrete structures typically show a high scatter and variability of achieved construction quality, however, it may be difficult to know which parts of the structure will have the highest rate of chloride ingress. At an early stage of operation, therefore, more locations for chloride control should be selected. After some time, when it becomes clearer where the highest rates of chloride ingress take place, further control of chloride ingress will be concentrated in these locations.

Normally, the rate of chloride ingress is faster at an early stage of operation than later on. The control of chloride ingress should therefore be carried out more frequently at an early stage. The first control should preferably be carried out shortly after completion of the structure in order to provide a proper reference level for the chloride content, while the next control could be carried out after a service period of, e.g., 10 years. Later on, both frequency and extent of control measurements will depend on the observed rates of chloride ingress.

At an early stage of operation, the control of chloride ingress can be carried out in a simpler way, just to get a rough indication of how fast the chloride ingress into the various parts of the structure takes place. From each location, such measurements can then be based on simple dust sampling from, for example, 5 × 16 mm drilling holes in 5 mm steps of depth. At a later stage, when a deeper chloride ingress is observed, however, the measurements from each location must be based on one or more concrete cores, from which very thin layers of dust samples are ground off and analyzed for a more detailed observation of the chloride ingress.

The chloride content can be measured by different procedures and test methods, as described in several recommendations and standards. Since field methods generally show a poorer accuracy, however, such measurements should only be used at an early stage of condition assessment in order to find out how fast the chloride ingress takes place. At a later stage, when it becomes necessary to obtain a more detailed distribution of the chloride ingress, the chloride analyses must be based on more accurate laboratory methods. Also, for a best possible comparison of chloride data from one inspection period to another, the same procedure and test method for control of the chloride ingress should be used.

If any instrumentation based on embedded probes should be part of the given plan for control of chloride ingress, preparation for this must also be made at an early stage of the durability design. Such instrumentation may be appropriate for control of chloride ingress in special or critical parts of the structure that later on may not be accessible for manual control.

Already at an early stage in the development of offshore concrete platforms for the oil and gas explorations in the North Sea, the BML Probe was developed (Figure 8.1). From this probe, not only could the rate of chloride ingress be followed, but also information about the general corrosion conditions and possible rates of corrosion obtained. Later, several new versions of such corrosion probes have been developed for control of chloride ingress and corrosion conditions in new concrete structures (Figures 8.2 to 8.4).

Although embedded probes for automatic control of chloride ingress can provide valuable data on rates of chloride ingress in certain locations of the structure, experience has shown that a few embedded probes can never replace a more complete control of chloride ingress in large concrete structures. For concrete structures that typically show a high scatter and variability of achieved construction quality, it may be difficult to know where the highest rates of chloride ingress will take place. Also, since embedded probes primarily provide information about how fast the chloride front moves into the concrete, more detailed information about the chloride ingress curve is needed in order to calculate the corrosion probability.

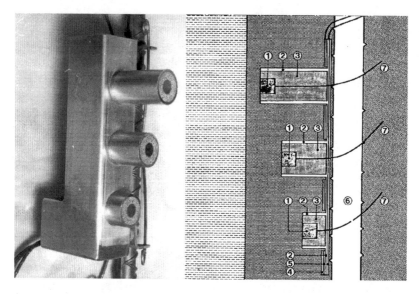

Figure 8.1 The BML Probe for automatic control of chloride ingress and corrosion conditions in concrete structures: (1) electrode, (2) casing of stainless steel, (3) insulation, (4) reference electrode, (5) insulation, (6) rebar system, and (7) shielded leads. (From Gjørv, O. E., and Vennesland, Ø., *Materials Performance*, 21(1), 33–35, 1982.)

8.3 PROBABILITY OF CORROSION

As soon as the chloride ingress becomes so deep that chloride data from at least six depths below the concrete surface can be determined, a regression analysis of all these chloride data for the given exposure period with curve fitting to Fick's second law of diffusion is carried out. As a result, a more complete chloride ingress curve than that shown in Figure 8.5 can be drawn. This provides the basis for determining some of the important durability parameters needed for calculating the corrosion probability.

Based on the above regression analysis, both the surface chloride concentration (C_S) and the apparent chloride diffusivity (D_a) are determined. A calculation of corrosion probability also needs information about the time dependence of the chloride diffusivity (α) and the concrete cover (X_C). Proper information about the concrete cover may either be available from the concrete quality control carried out during concrete construction or determined in other ways. At an early stage of operation, however, information about the time dependence of the chloride diffusivity (α) is not available. For an early calculation of corrosion probability, therefore, the calculation must be based on an estimated, empirical α-value for the given concrete in the given environment. As soon as two or more values for the

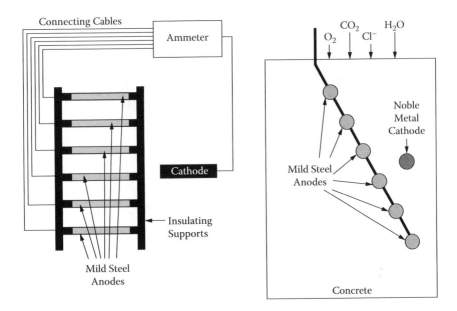

Figure 8.2 The anode ladder probe for automatic control of chloride ingress and corrosion conditions in concrete structures. (From Raupach, M., and Schiessl, P., *Construction and Building Materials*, 11, 207–214, 1997.)

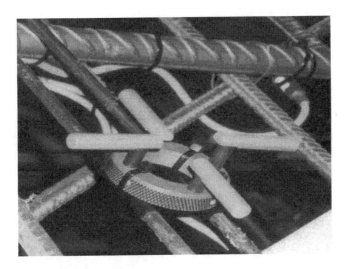

Figure 8.3 The CorroWatch Multisensor for automatic control of chloride ingress and corrosion conditions in concrete structures. (From Force Technology, CorroWatch Multiprobe, Force Technology, Brøndby, Denmark, 2008, http://www.forcetechnology.com.)

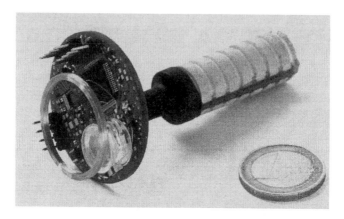

Figure 8.4 The Braunschweig Corrosion Sensor for automatic control of chloride ingress and corrosion conditions in concrete structures. (From Fraunhofer Institute, Braunschweig Corrosion Sensor, Fraunhofer Institute for Microelectronic Circuits and Systems, Duisburg, Germany, 2011, http://www.ims.fraunhofer.de.)

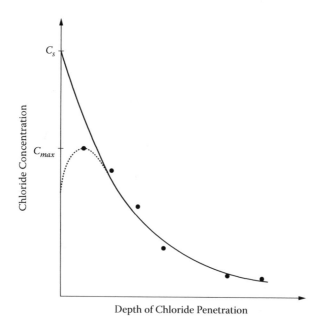

Figure 8.5 The result of a regression analysis of all observed data on chloride ingress for a given period of exposure.

apparent chloride diffusivity (D_a) become available from further control measurements, however, a more accurate α-value can be determined, as shown in the following (Poulsen and Mejlbro, 2006):

$$D_{a2} = D_{a1} \cdot \left(\frac{t_1}{t_2} \right)^{\alpha} \tag{8.1}$$

where D_{a1} and D_{a2} are the apparent chloride diffusivities after exposure periods of t_1 and t_2, respectively. The time dependence of the chloride diffusivity (α) then becomes

$$\alpha = \frac{\ln \left(\dfrac{D_{a(t_2)}}{D_{a(t_1)}} \right)}{\ln \left(\dfrac{t_1}{t_2} \right)} \tag{8.2}$$

As soon as three or more values for the apparent chloride diffusivity become available, even more accurate α-values can be determined based on a regression analysis, and the more future values for the apparent chloride diffusivity that become available, the more accurate the α-value will be. Hence, the calculation of corrosion probability will gradually be more accurate during operation of the structure. Before the probability of corrosion becomes too high, however, appropriate protective measures for control of further chloride ingress must be implemented.

8.4 PROTECTIVE MEASURES

Depending on type of protective measure, the observed rates of chloride ingress can be either reduced or completely stopped (Chapter 5). If the chlorides have not reached too deep through the concrete cover, a proper surface treatment or coating may slow down the further rate of chloride ingress. If the chlorides have already reached too deep, however, cathodic prevention is the only protective measure that can stop the further chloride ingress, and thus avoid the development of steel corrosion.

In this way, proper control of chloride ingress, in combination with calculations of corrosion probability, provides the basis for regular condition assessment and preventive maintenance of the structure. If the chlorides are allowed to reach the embedded steel and corrosion starts, it is only a question of time before visual damage develops and necessary repairs have to be carried out.

8.5 REPAIRS

Of the various types of repairs of chloride-induced corrosion, patch repairs are not very effective in bringing the corrosion under control. As was discussed in Chapter 2, the poor effect of patch repairs was first observed on the San Mateo–Hayward Bridge (1929) in the Bay Area of San Francisco already in the early 1950s (Gewertz et al., 1958). This bridge had been extensively patch repaired, first by local cleaning of the damaged areas and then by shotcreting, but after a short period of time, continued steel corrosion was observed. During his extensive field investigation of this bridge, Stratfull (1974) for the first time carried out detailed potential mapping of the patch-repaired structure by using half-cell copper-copper sulfate electrodes. In this way, he demonstrated that the patched areas had typically formed cathodic areas, while anodic areas had formed adjacent to the patched areas, as shown in Figure 8.6.

In the early 1970s, Stratfull also demonstrated that cathodic protection would be the most efficient way to get chloride-induced corrosion under control (Stratfull, 1974). Later on, extensive experience has confirmed that cathodic protection is the only way of getting heavy chloride-induced steel corrosion under control (Broomfield, 1997; Bertolini et al., 2004).

8.6 CASE STUDY

8.6.1 General

In order to demonstrate how a condition assessment of a marine concrete harbor structure may be carried out, the cruise terminal in Trondheim Harbor shown in Figure 2.20 (Chapter 2) was selected as a case study. The condition assessment of this structure was carried out after eight years of operation, and no control of chloride ingress had previously been carried out. The structure, which was built in 1993, has a water front of 95 m and consists of an open concrete deck on top of driven steel tubes filled with concrete. The deck is a beam and slab type of deck, the top of which has an elevation of 3 m above mean water level. The specified durability was based on the then-current Norwegian Concrete Codes with a water/binder ratio of ≤ 0.45, a binder content of ≥ 300 kg/m^3, and a minimum concrete cover of 50 mm. The applied concrete, which had a 28-day compressive strength of 45 MPa, was produced with 380 kg/m^3 high-performance portland cement and 19 kg/m^3 silica fume (5%). According to the files from the concrete construction period, all requirements to both concrete quality and concrete cover had been fulfilled.

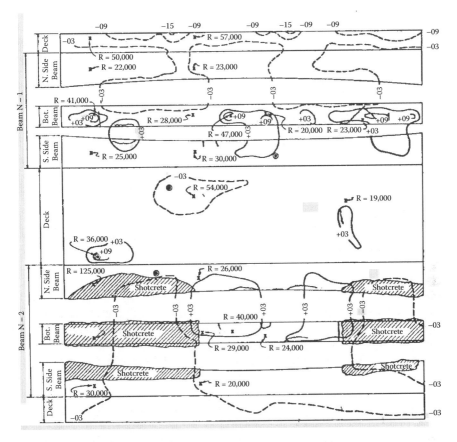

Figure 8.6 Equipotential contours demonstrating how the local patch repairs by shotcreting had typically formed a pattern of anodic areas (solid lines) and cathodic areas (dotted lines) along the concrete surface of the San Mateo–Hayward Bridge (1929). (From Gewertz, M. W. et al., *Causes and Repair of Deterioration to a California Bridge due to Corrosion of Reinforcing Steel in a Marine Environment*, Highway Research Board Bulletin 182, National Research Council Publication 546, National Academy of Sciences, Washington, DC, 1958.)

8.6.2 Condition assessment

During the condition assessment, no visual damage was revealed, but more thorough field investigations were carried out, including electrochemical surface potential mapping and a number of measurements of both chloride ingress and concrete cover. For assessment of the in situ quality of the concrete, a few concrete cores from the deck slab were also removed for determination of the chloride diffusivity (rapid chloride migration (RCM)).

Table 8.1 Input parameters for the durability analyses of the two investigated deck beams

Deck beam	B1	B2
Chloride loading, C_S (% by wt. of cement)	N(3.2, 0.74)	N(4.1, 0.17)
Temperature, T (°C)	10	
Chloride diffusivity, D_a (×10^{-12} m²/s)	N(0.95, 0.17)	N(1.1, 0.27)
Time dependence factor, α	N(0.4, 0.07)	
Age of concrete, t_0 (years)	8	
Critical chloride content, C_{CR} (% by wt. of cement)	N(0.4, 0.01)	
Concrete cover, X_c (mm)	N(50.0, 10.0)	N(48.7, 5.5)
Service period (years)	100	

The electrochemical surface potential mapping revealed that in some parts of the most exposed deck beams, a depassivation with an early stage of steel corrosion had already taken place. For most of the concrete deck, however, the chloride ingress had not yet reached the embedded steel.

As part of the overall condition assessment, two of the most exposed deck beams were selected for a more detailed investigation and subjected to durability analyses, the input parameters for which are shown in Table 8.1. For both deck beams, a deep chloride ingress was observed, but the chloride ingress was somewhat deeper in deck beam B2 than in deck beam B1, giving somewhat different values for the chloride loading (C_S). Values for the apparent chloride diffusivity (D_a) were also obtained from the observed chloride ingress curves, while the values for concrete cover (X_C) were based on a number of cover measurements. As input parameter for the time dependence factor (α), however, only an empirical value based on current experience from similar concrete structures in similar environments was adopted. For both deck beams, Figure 8.7 demonstrates that a 10% probability of corrosion had already been reached for a 10-year service period. For deck beam B2, the electrochemical surface potentials revealed that depassivation had already taken place.

Although the specified durability requirements with respect to water/binder ratio, binder content, and concrete cover had been fulfilled, the condition assessment revealed that steel corrosion had already started in the most exposed parts of the structure after a service period of about eight years. Based on the RCM testing of three Ø100 mm concrete cores from the deck slab, an average chloride diffusivity of 10.7 × 12^{-12} m²/s with a standard deviation of 1.1 was observed. Such a level of chloride diffusivity after eight years of curing in a moist environment indicates that the concrete showed a low to moderate resistance to chloride ingress (Table 4.2).

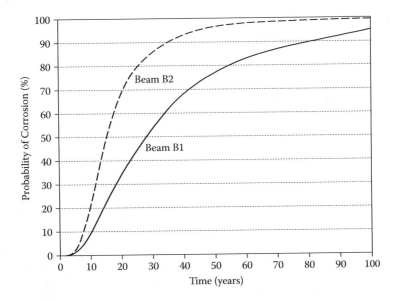

Figure 8.7 Probability of corrosion in the two deck beams after eight years of operation.

8.6.3 Protective measure

Since the early stage of steel corrosion had not yet caused any visible damage, the application of a cathodic prevention system would have been a most appropriate measure in order to stop both the already ongoing corrosion and the further ingress of chlorides. However, since the electrical continuity within the reinforcement system was very poor and the application of a cathodic prevention system would therefore have been quite expensive, the owner decided only to apply a hydrophobic surface treatment underneath the concrete deck. At least, it was believed that such a protective measure would retard the further chloride ingress into those parts of the concrete deck where the observed chloride ingress had not yet reached too deep.

In order to investigate the effect of the applied hydrophobic surface treatment later on, some control measurements of the chloride ingress after two years of further exposure were carried out. As can be seen from Figure 8.8, a certain redistribution of the chloride content had typically taken place, showing a reduced chloride content in the surface layer of the concrete cover with a deeper ingress of the chloride front further in. Such a redistribution of the chloride ingress has also previously been reported in the literature (Arntsen, 2001). Although the rate of chloride ingress was generally retarded, the surface treatment had not been very effective in preventing the

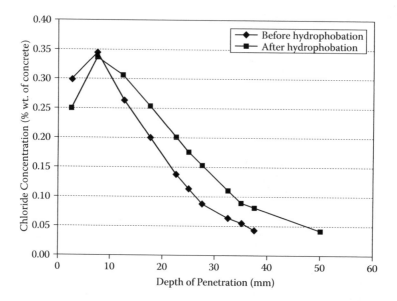

Figure 8.8 Chloride ingress before hydrophobic surface treatment of the concrete deck and after two years of further exposure. (From Årskog et al., Effect of Surface Hydrophobation on Chloride Penetration into Concrete Harbor Structures, in *Proceedings, Fourth International Conference on Concrete under Severe Conditions—Environment and Loading,* vol. I, ed. B. H. Oh et al., Seoul University and Korea Concrete Institute, Seoul, South Korea, 2004, pp. 441–448.)

further ingress of chlorides. Therefore, such a protective measure should not be applied if the chlorides have already reached too deep.

REFERENCES

Arntsen, B. (2001). *In-Situ* Experiences on Chloride Redistribution in Surface-Treated Concrete Structures. In *Proceedings, Third International Conference on Concrete under Severe Conditions—Environment and Loading,* vol. 1, ed. N. Banthia, K. Sakai, and O.E. Gjørv. University of British Columbia, Vancouver, pp. 95–103.

Årskog, V., Liu, G., Ferreira, M., and Gjørv, O.E. (2004). Effect of Surface Hydrophobation on Chloride Penetration into Concrete Harbor Structures. In *Proceedings, Fourth International Conference on Concrete under Severe Conditions—Environment and Loading,* vol. 1, ed. B.H. Oh, K. Sakai, O.E. Gjørv, and N. Banthia. Seoul National University and Korea Concrete Institute, Seoul, pp. 441–448.

Bertolini, L., Elsener, B., Pediferri, P., and Polder, R. (2004). *Corrosion of Steel in Concrete—Prevention, Diagnosis, Repair.* Wiley-VCH, Weinheim, Germany.

BRIME. (1997). *Bridge Management in Europe*, EC-DG-VII-RTD, European Union—Program Contract RO-97-SC.2220.

Broomfield, J.P. (1997). *Corrosion of Steel in Concrete*. E & FN Spon, London.

Eri, J., Vælitalo, S.H., Gjørv, O.E., and Pruckner, F. (1998). Automatic Monitoring for Control of Steel Corrosion in Concrete Structures. In *Proceedings, Second International Conference on Concrete under Severe Conditions—Environment and Loading*, vol. 2, ed. O.E. Gjørv, K. Saka, and N. Banthia. E & FN Spon, London, pp. 1007–1015.

Force Technology. (2008). CorroWatch Multiprobe. Force Technology, Brøndby, Denmark, www.forcetechnology.com.

Fraunhofer Institute. (2011). Braunschweig Corrosion Sensor. Fraunhofer Institute for Microelectronic Circuits and Systems, Duisburg, Germany, www.ims.fraunhofer.de.

Gewertz, M.W., Tremper, B., Beaton, J.L., and Stratfull, R.F. (1958). *Causes and Repair of Deterioration to a California Bridge due to Corrosion of Reinforcing Steel in a Marine Environment*, Highway Research Board Bulletin 182, National Research Council Publication 546. National Academy of Sciences, Washington, DC.

Gjørv, O.E. (2002). Durability and Service Life of Concrete Structures. In *Proceedings, First FIB Congress 2002*, session 8, vol. 6. Japan Prestressed Concrete Engineering Association, Tokyo, pp. 1–16.

Gjørv, O.E., and Vennesland, Ø. (1982). A New Probe for Monitoring Steel Corrosion in Offshore Concrete Platforms. *Materials Performance*, 21(1), 33–35.

Grigg, N.S. (1988). *Infrastructure Engineering and Management*. John Wiley & Sons, New York.

Hudson, S.W., Carmichael, R.F., Moser, L.O., Hudson, W.R., and Wilkes, W.J. (1987). *Bridge Management Systems*, NCHRP Report 300. Transportation Research Board, National Research Council, Washington, DC.

O'Connor, D.S., and Hyman W.A. (1989). *Bridge Management Systems*, Report FHWA-DP-71-01R, Demonstration Project 71. Demonstration Projects Division, Federal Highway Administration, Washington, DC.

Poulsen, E., and Mejlbro, L. (2006). *Diffusion of Chlorides in Concrete—Theory and Application*. Taylor & Francis, London.

Raupach, M., and Schiessl, P. (1997). Monitoring System for the Penetration of Chlorides, Carbonation and the Corrosion Risk for the Reinforcement, *Construction and Building Materials*, 11, 207–214.

RIMES. (1997). *Road Infrastructure Maintenance Evaluation Study on Pavement and Structure Management System*, EC-DG-VII-RTD, Programme-Contract RO-97-SC 1085/1189.

Stratfull, R.F. (1974). *Experimental Cathodic Protection of a Bridge*, Research Report 635117-4, FHWA D-3-12, Department of Transportation, Sacramento, CA.

Tromposch, E.W., Dunaszegi, L., Gjørv, O.E., and Langley, W.S. (1998). Northumberland Strait Bridge Project—Strategy for Corrosion Protection. In *Proceedings, Second International Conference on Concrete under Severe Conditons—Environment and Loading*, vol. 3, ed. O.E. Gjørv, K. Sakai, and N. Banthia. E & FN Spon, London, pp. 1714–1720.

Chapter 9

Practical applications

9.1 GENERAL

In recent years, a number of new important concrete infrastructures along the Norwegian coastline have been constructed, and for most of them, the specified durability has mainly been based on the durability requirements according to the current concrete codes, with a water/binder ratio of ≤0.40 and a binder content of ≥330 kg/m³ (Standard Norway, 2003a, 2003b, 2003c). For new concrete coastal bridges, a water/binder ratio of ≤0.38 has been specified (NPRA, 1996).

In order to obtain some further information about the quality of the concrete typically being applied to all the above structures, samples of the concrete from some of these construction sites were collected in order to test the development of chloride diffusivity based on the rapid chloride migration (RCM) method (NORDTEST, 1999). Although all of these types of concrete fulfilled the specified durability requirements for a 100-year service life in a severe marine environment, it may be seen from Table 9.1 that the chloride diffusivity or the resistance to chloride ingress of the various types of concrete varied within wide limits. All this testing was carried out on concrete specimens that were separately cast from the various construction sites and then water cured in the laboratory until time of testing.

In order to obtain an increased and more controlled durability and service life, some of the above concrete structures were also subjected to a durability design and concrete quality control according to the NAHE recommendations of 2004 (NAHE, 2004a, 2004b, 2004c). In the following, a brief outline of the experience gained with these practical applications is given, the procedures for which are practically the same as that described in the previous chapters. Also, the DURACON software (http://www.pianc.no/duracon/php) for calculations of corrosion probability is the same.

All the following concrete structures were built in the harbor region of Oslo City in recent years. One of the structures was the first part of a new container terminal completed in 2007 (Container Terminal 2), while the other structures were part of a more comprehensive project (New City

Table 9.1 Chloride diffusivity (RCM) of concrete typically being applied for new marine concrete construction along the Norwegian coastline

| | Chloride diffusivity ($\times 10^{-12}$ m²/s) | | | | | | | | | |
| | Age (days) | | | | | | | | | |
Construction site	14	28	60	90	180	365	400	460	620	730
Container Terminal 1, Oslo (2002)	13.5	6.0	4.4	3.8	3.0	—	—	—	—	—
Gas Terminal, Aukra (2005)	17.6	6.8	4.3	2.3	—	—	1.5	—	—	—
Eiksund Bridge, Eiksund (2005)	14.1	4.4	3.8	3.4	3.1	—	—	3.0	—	—
Container Terminal 2, Oslo (2007)	14.0	6.9	4.6	2.4	1.2	0.7	—	—	—	0.7
New City Development, Oslo (2010)	4.7	1.6	0.4	0.4	0.3	0.2	—	—	0.2	0.2

Development) completed in 2010. As a reference project, a brief outline of experience gained from another container terminal completed in 2002 (Container Terminal 1) is also included, for which the durability requirements were only based on the current concrete codes.

9.2 CONTAINER TERMINAL I, OSLO (2002)

9.2.1 Specified durability

This concrete harbor structure consists of an open beam and slab type of deck on top of driven steel tubes filled with concrete (Figure 9.1). The structure, which has a waterfront of 144 m, was constructed in two steps and completed in 2002. At that time, the above procedures for durability design were not available. The specifications for a 100-year service life were therefore based on the then-current Norwegian Concrete Codes with some additional requirements:

- $W/(C + k \cdot S)$: 0.40 ± 0.03
- Minimum cement content (C): 370 kg/m³
- Silica fume (S): 6–8% by weight of cement
- Air content: $5.0 \pm 1.5\%$

For the above water/binder ratio, k is an empirical efficiency factor of two for the use of silica fume. A nominal concrete cover to the structural steel of 75 ± 15 mm was also specified.

Figure 9.1 Container Terminal I (2002) in Oslo Harbor consists of an open concrete deck on top of driven steel tubes filled with concrete.

Although no probability-based durability design was carried out, Oslo Harbor KF, as the owner of the structure, required a best possible documentation of the achieved construction quality during the construction period. A type of concrete quality control and documentation similar to that described in Chapters 6 and 7 was carried out.

9.2.2 Achieved construction quality

In order to provide documentation of the achieved construction quality, a number of Ø100 × 200 mm concrete cylinders were produced at an early stage of concrete construction and tested for chloride diffusivity (RCM) at different periods of curing. All these specimens were water cured in the laboratory at 20°C until time of testing for a period of up to half a year. In addition, a representative dummy element was also produced on the construction site. Both from the given structure under construction and the dummy element, a number of Ø100 mm concrete cores were removed during the construction period for testing the in situ development of chloride diffusivity (Figure 9.2). In addition, regular control of the achieved concrete cover was carried out, which included 153 individual measurements showing an average value of 65 mm with a standard deviation of 7 mm.

Based on the separately cast and water-cured concrete specimens in the laboratory, a 28-day chloride diffusivity (D_{28}) of 6.0×10^{-12} m²/s was obtained. Combined with the achieved concrete cover from the quality control, a durability analysis was carried out in order to calculate the service period before 10% probability of corrosion would be reached. All the necessary input parameters to this durability analysis are shown in Table 9.2. The chloride loading (C_S) was based on long-term experience from similar

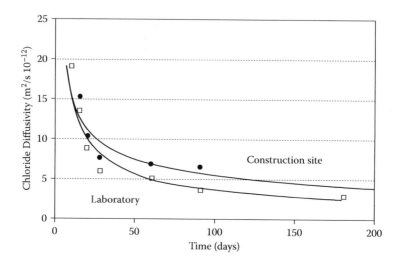

Figure 9.2 Development of chloride diffusivity both on the construction site and in the laboratory.

Table 9.2 Input parameters for analysis of achieved durability

Input parameter	
Chloride loading, C_S (% by wt. of binder)	$N^a(3.8, 0.9)$
Age of concrete at exposure, t (days)	28
Temperature, T (°C)	10
Chloride diffusivity, D_{28} ($\times 10^{-12}$ m²/s)	$N(6.0, 0.6)$
Age of concrete at testing, t (days)	28
Time dependence factor, α	$N(0.4, 0.08)$
Critical chloride content, C_{CR} (% by wt. of binder)	$N(0.4, 0.08)$
Concrete cover, X_c (mm)	$N(65, 7)$
Service period (years)	100

a Normal distribution with average value and standard deviation.

concrete structures in Oslo Harbor after exposure periods of up to 80 years. As part of the environmental loading, an annual average temperature (T) of 10°C and a 28-day age for chloride exposure (t') were selected. Since the given concrete was based on a portland cement (CEM I 52.5 LA) in combination with 6% silica fume by weight of cement, an empirical time dependence factor for the chloride diffusivity (α) of 0.40 was selected, together with a critical chloride content (C_{CR}) of 0.4% by weight of binder. As can be seen from Figure 9.3, the 10% probability of corrosion would be reached

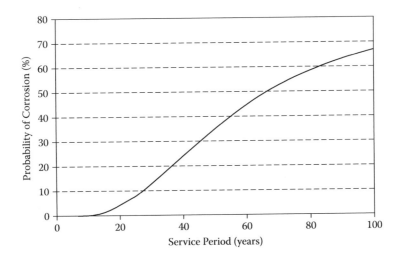

Figure 9.3 Probability of corrosion versus time for Container Terminal I in Oslo Harbor (2002).

after a service period of approximately 30 years, while the probability of corrosion after 100 years would be approximately 65%.

9.2.3 In situ quality

Based on the achieved chloride diffusivity after 200 days on the construction site of 3.9×10^{-12} m²/s, in combination with the control data on achieved concrete cover, a new durability analysis was carried out. With all the other input parameters kept the same as that shown in Table 9.2, the new analysis showed that a probability for corrosion of approximately 80% during a service period of 100 years would be reached.

9.2.4 Potential quality

Based on the water-cured concrete specimens in the laboratory for a period of up to 180 days, a chloride diffusivity of 3.0×10^{-12} m²/s was obtained. Combined with the achieved data on concrete cover, a final durability analysis was carried out in order to provide documentation of the potential construction quality of the structure. With all the other input parameters kept the same (Table 9.2), the new analysis showed that a probability of corrosion of approximately 60% during a service period of 100 years would be reached.

Based on all the above results for achieved construction quality, general experience indicates that a relatively high level of maintenance costs may be expected for maintaining a proper long-term performance and service-ability of the given concrete structure during the specified service life of 100 years.

9.3 CONTAINER TERMINAL 2, OSLO (2007)

9.3.1 General

From 2005 to 2007, the first part of another container terminal in Oslo Harbor was constructed. This structure, having a waterfront of about 300 m, also consists of an open concrete deck on driven steel tubes filled with concrete. According to the current Norwegian Concrete Codes for a 100-year service life, the minimum durability requirements should include a water/binder ratio of ≤0.40, a binder content of ≥330 kg/m³ and a minimum concrete cover of 60 mm (Standard Norway, 2003a, 2003b, 2003c). A total air content of 4 to 6% should also provide proper frost resistance.

In order to obtain an increased and more controlled durability of the given structure, however, Oslo Harbor KF, as the owner of the structure, decided to apply the new NAHE recommendations of 2004, which had just been published (NAHE, 2004a, 2004b, 2004c).

9.3.2 Specified durability

As an overall durability requirement to the given structure, the owner required a service period of 100 years before 10% probability of corrosion would be reached. In order to satisfy this durability requirement, an initial durability analysis with input parameters as shown in Table 9.3 was carried out. This analysis provided the basis for establishing the necessary requirements to concrete quality and concrete cover. With a location very close

Table 9.3 Input parameters for the initial durability analysis

Input parameter	
Chloride load, C_S (% by wt. of binder)	$N^a(3.8, 0.9)$
Age of concrete at exposure, t' (days)	28
Temperature, T (°C)	10
Chloride diffusivity, D_{28} (×10⁻¹² m²/s)	$N(5.0, 1.0)$
Age of concrete at testing, t (days)	28
Time dependence factor, α	$N(0.6, 0.12)$
Critical chloride content, C_{CR} (% by wt. of binder)	$N(0.4, 0.08)$
Concrete cover, X_C (mm)	$N(90, 11)$
Service period (years)	100

a   Normal distribution with average value and standard deviation.

to the previous structure (Container Terminal 1), the same data for environmental loading as those selected for this structure were adopted. Since previous experience also had shown that the risk for early-age exposure during concrete construction was high, the time for chloride exposure (t') of 28 days was also kept the same. Based on current experience for the use of fly ash cement, a concrete quality with a 28-day chloride diffusivity of 5.0×10^{-12} m²/s, a time dependence factor (α) of 0.60, and a critical chloride content (C_{CR}) of 0.4% were further selected.

Based on the above durability analysis, it was shown that a chloride diffusivity of $D_{28} \leq 5.0 \times 10^{-12}$ m²/s in combination with a concrete cover of 90 ± 15 mm would satisfy the overall durability requirement with a proper margin. These data were therefore adopted as a basis for specifying the necessary requirements to chloride diffusivity and concrete cover.

In order to establish a concrete mixture that would meet the specified chloride diffusivity, some preliminary tests were carried out by the contractor. Based on a pure portland cement (CEM I 52.5 LA) and 4% silica fume by weight of cement, three different trial mixtures with 20, 40, and 60% replacements of the portland cement by a low-calcium-containing fly ash (FA) were produced and tested. Based on these tests, the mixture based on 60% FA was selected by the contractor for the further concrete work. Although this type of concrete showed a somewhat higher 28-day value of chloride diffusivity (D_{28}) than that specified, the further development of the chloride diffusivity showed very good results. Since the documentation of frost resistance of this concrete was also very good, this type of concrete was accepted for further concrete production.

9.3.3 Compliance with specified durability

Shortly after start of concrete construction, a representative dummy element in the form of a concrete slab with dimensions of $2.0 \times 2.0 \times 0.5$ m without any reinforcement was produced on the construction site. From the same concrete batch as that used for the dummy element, a number of 100 mm concrete cubes and Ø100 × 200 mm concrete cylinders were also produced on the construction site. Already, the next day, all of these test specimens were sent to the laboratory for testing and establishing the necessary calibration curve. The electrical resistivity measurements were carried out by use of a Wenner device, while the parallel testing of chloride diffusivity was based on the RCM method. With parallel measurements of chloride diffusivity and electrical resistivity after curing periods of approximately 14, 28, 56, and 90 days (Table 9.4), a calibration curve as shown in Figure 9.4 was established. This calibration curve was later used for the indirect quality control of the chloride diffusivity (D_{28}) based on electrical resistivity measurements on all the 100 mm concrete cubes used for the regular control of the 28-day compressive strength during concrete construction.

Table 9.4 Test results for establishing the necessary calibration curve

Testing age (days)	Chloride diffusivity[a] ($\times 10^{-12}$ m²/s)	Electrical resistivity[a] (ohm·m)
14	14.0, 1.9	146, 9
28	6.9, 0.1	307, 21
56	4.6, 0.6	508, 41
90	2.4, 0.5	699, 131

[a] Average value and standard deviation.

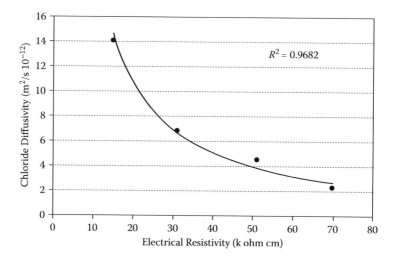

Figure 9.4 Calibration curve for quality control of chloride diffusivity based on electrical resistivity measurements.

During concrete construction, a total of 344 individual measurements of electrical resistivity were carried out, and this quality control showed an acceptable level of chloride diffusivity without large deviations throughout the concrete construction period. As a result, an average value of 7.9 × 10^{-12} m²/s for the chloride diffusivity with a standard deviation of 3.2 was obtained (Table 9.5).

Throughout concrete construction, a number of control measurements of achieved concrete cover were also carried out. Due to the very thick concrete cover and very congested reinforcement (Figures 9.5 and 9.6), it was not easy to obtain reliable measurements based on a conventional cover meter. Therefore, all control measurements were based on manually observed concrete covers on protruding bars in the construction joints during concrete construction. Based on altogether 68 individual measurements, an average concrete cover of 99 mm with a standard deviation of 11 mm was obtained.

Table 9.5 Obtained chloride diffusivity (D_{28}) based on electrical resistivity measurements during concrete construction

Testing age (days)	Electrical resistivity[a] (ohm·m)	Chloride diffusivity[a] ($\times 10^{-12}$ m²/s)
28	260, 102	7.9, 3.2

[a] Average value and standard deviation.

Figure 9.5 Protruding bars in the construction joints of the concrete deck.

Based on the above control data on both chloride diffusivity and concrete cover as new input parameters, a new durability analysis was carried out with all the other originally selected input parameters kept the same (Table 9.6). As a result, a probability for corrosion of approximately 2% over a 100-year service period was obtained, showing that the specified durability was achieved with a proper margin.

Figure 9.6 The concrete deck had very congested reinforcement.

Table 9.6 Input parameters for the control of compliance with specified durability

Input parameter	
Chloride loading, C_S (% by wt. of binder)	$N^a(3.8, 0.9)$
Age of concrete at exposure, t' (days)	28
Temperature, T (°C)	10
Chloride diffusivity, D_{28} ($\times 10^{-12}$ m²/s)	$N(7.9, 3.2)$
Age of concrete at testing, t (days)	28
Time dependence factor, α	$N(0.6, 0.12)$
Critical chloride content, C_{CR} (% by wt. of binder)	$N(0.4, 0.08)$
Concrete cover, X_C (mm)	$N(99, 11)$
Service period (years)	100

[a] Normal distribution with average value and standard deviation.

9.3.4 In situ quality

During the construction period, a number of Ø100 mm concrete cores were removed at various stages from both the given structure under construction and the separately produced dummy element on the construction site. Immediately upon removal, all these cores were properly wrapped in plastic and sent to the laboratory for testing of the chloride diffusivity, the results of which, up to an age of one year, are shown in Figure 9.7. In the same figure, the development of chloride diffusivity based on water-cured specimens in the laboratory is also shown.

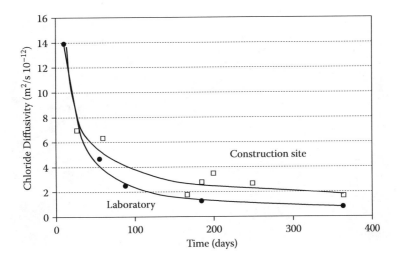

Figure 9.7 Development of chloride diffusivity on the construction site and in the laboratory.

Based on the achieved chloride diffusivity on the construction site after one year of 1.5×10^{-12} m²/s with a standard deviation of 0.54, in combination with the achieved concrete cover, a new durability analysis was carried out. As a result, a probability for corrosion of approximately 0.05% after a service period of 100 years was obtained. Such a result indicates that the achieved construction quality on the construction site during the construction period was very good.

9.3.5 Potential quality

Based on the achieved chloride diffusivity after one year of the water-cured specimens in the laboratory at 20°C, in combination with the achieved concrete cover, a further durability analysis was carried out. For an obtained chloride diffusivity of 0.7×10^{-12} m²/s with a standard deviation of 0.02, this durability analysis showed a probability for corrosion of <0,001%, which was hardly detectable after a service period of 100 years. This result indicates that the potential construction quality of the given concrete structure was extremely good.

9.4 NEW CITY DEVELOPMENT, OSLO (2010)

9.4.1 General

In 2005, the New City Development on Tjuvholmen in the harbor region of Oslo City began. This development project includes a number of sea-spaced

Figure 9.8 The New City Development on Tjuvholmen in the harbor region of Oslo City. (Photo by Terje Løchen.)

concrete substructures in water depths of up to 20 m, on top of which a number of business and apartment buildings have been built (Figure 9.8). All the concrete substructures were finished by 2010, most of which include large underwater parking areas. In the shallower water, the structures include a solid concrete bottom slab on the seabed, surrounded by external concrete walls partly protected by riprap or wooden cladding and partly freely exposed to the tides. In the deeper water, some structures include an open concrete deck on columns of driven steel pipes filled with concrete. In the deepest water, four large concrete caissons were prefabricated in dry dock, moved into position, and submerged in water up to 20 m deep (Figure 9.9a and b). Three of these structures provide up to four levels of submerged parking (Figure 9.10).

For all the concrete substructures, the owner and developer of the project required a service life of 300 years, which means that a highest possible durability and long-term performance of the given structures were required. As minimum durability requirements to the structures, all requirements in the current Norwegian Concrete Codes for a 100-year service life had to be fulfilled (Standard Norway, 2003a, 2003b, 2003c).

In order to obtain an increased and more controlled durability of all the concrete substructures for the first four parts of the project, these structures were subjected to a probability-based durability design according to the NAHE recommendations of 2004 (NAHE, 2004a, 2004b, 2004c). All of these concrete structures, which were produced by one contractor (Contractor A), mainly included solid concrete bottom slabs on the seabed surrounded by external concrete walls exposed to the tidal zone. All the

(a)   (b)

Figure 9.9 Large concrete caissons were prefabricated in a dry dock, moved into position, and submerged in water up to 20 m deep.

Figure 9.10 Section showing how large, prefabricated concrete caissons after installation provide up to four levels of submerged parking.

other concrete substructures in the last four parts of the project were produced by another contractor (Contractor B), and these structures mainly included four large caissons prefabricated in a dry dock at two different construction sites. In addition, a number of open concrete decks were also produced, partly as prefabricated elements, but mostly produced in situ. For all these concrete structures, the contracts were primarily based on the durability requirements for a 100-year service life according to the current Norwegian Concrete Codes with some additional durability requirements.

For all the concrete structures throughout the whole project, however, the owner and developer also required that a performance-based concrete quality control with documentation of achieved construction quality according to the current NAHE recommendations be carried out. Thus, a unique opportunity occurred to compare the results and experience obtained by use of performance- versus prescriptive-based durability requirements; this is briefly outlined and discussed in the following.

9.4.2 Specified durability

9.4.2.1 Performance-based durability requirements

Since the current procedures for probability-based durability design are not considered valid for a service period of more than 150 years, the overall durability requirement for all the concrete structures in the first four parts of the project was based on a probability of corrosion as low as possible and not exceeding 10% for a 150-year service period. In order to further increase and ensure the durability, an additional protective measure also had to be applied, which for the first structure was provisions for future cathodic prevention in combination with embedded probes for chloride monitoring. For all the other concrete structures in the first four parts of the project, however, the additional protective measure was a partial replacement of the black steel with stainless steel reinforcement (W.1.4301).

As a basis for selecting a proper combination of concrete quality and concrete cover that would meet the above durability requirement, an initial durability analysis was carried out based on current experience with the chloride diffusivity of different types of concrete (Chapter 3). On this basis, a concrete based on blast furnace slag cement with 70% slag (CEM III/B 42.5 LH HS) in combination with 10% silica fume typically giving a 28-day chloride diffusivity (D_{28}) of 2.0×10^{-12} m²/s was adopted. A nominal concrete cover of 100 ± 10 mm was also adopted, while all the other input parameters needed for the durability analysis were based on typical data for the marine environment in Oslo Harbor (Table 9.7). As a result, a probability for corrosion of less than 0.3% after a 150-year service period for the most exposed parts of the structures would be obtained. Therefore, the above values for both the 28-day chloride diffusivity and the nominal concrete cover were adopted as intended values for the first concrete structure. A proper frost resistance of the concrete was also required, and in order to reduce the risk for early-age cracking of the 100 mm concrete cover, a certain dosage of synthetic fibers was added to the concrete.

While provisions for future cathodic prevention were applied for all the exposed walls in the first concrete structure, no additional protective measure for the continuously submerged bottom slab was considered necessary due to the very low oxygen availability in this submerged part of the structure.

Table 9.7 Input parameters for the initial durability analysis

Input parameter	
Chloride loading, C_S (% by wt. of binder)	$N^a(3.8, 0.9)$
Age of concrete at exposure, t' (days)	28
Temperature, T (°C)	10
Chloride diffusivity, D_{28} ($\times 10^{-12}$ m^2/s)	$N(2.0, 0.4)$
Age of concrete at testing, t (days)	28
Time dependence factor, α	$N(0.5, 0.1)$
Critical chloride content, C_{CR} (% by wt. of binder)	$N(0.4, 0.08)$
Concrete cover, X_C (mm)	$N(100, 7)$
Service period (years)	150

[a] Normal distribution with average value and standard deviation.

For the second concrete structure, which consisted of an open concrete deck on columns of driven steel pipes filled with concrete, an additional protective measure based on partial use of stainless steel (W.1.4301) was applied. Since this protective measure very soon showed to be a much more simple and robust technical solution, such a protective measure was also selected and applied for the most exposed parts of all the further concrete structures in the first four parts of the project.

When the black steel was replaced by stainless steel in the outer layer of the rebar system, the effective concrete cover to the black steel further in increased to more than 150 mm. As a consequence, the nominal concrete cover to the stainless steel could be reduced to 85 ± 10 mm, but still, a very low probability of corrosion would be maintained. At the same time, any addition of fibers to the concrete for these parts of the structures was no longer necessary. For all the solid bottom slabs, however, both black steel reinforcement with a nominal concrete cover of 100 ± 10 mm and concrete with synthetic fibers were still applied.

9.4.2.2 Prescriptive-based durability requirements

For all the concrete substructures in the last four parts of the project (Contractor B), the durability requirements were primarily based on the prescriptive durability requirements according to the current Norwegian Concrete Codes for a 100-year service life, including a water/binder ratio of ≤0.40 and a minimum binder content of 330 kg/m^3. These requirements also include nominal concrete covers for the permanently submerged parts and the tidal and splash zones of 60 and 70 mm, respectively. In order to increase the durability of the given structures, however, the nominal concrete cover for the permanently submerged concrete slabs of the caissons was increased from 60 to 80 mm, while for all the external walls with tidal and splash exposure it was increased from 70 to 100 mm. For the submerged parts of the structures, cathodic prevention in the form of sacrificial anodes was also applied, but above water, only provisions for future installation of cathodic prevention in combination with embedded instrumentation for chloride monitoring were applied. For those parts of the structures that would be exposed to freezing and thawing, an air content of 4 to 6% was also required in order to ensure proper frost resistance.

9.4.3 Concrete quality control

For all the concrete structures throughout the whole project, the owner and developer of the project required that a performance-based concrete quality control with documentation of achieved construction quality according to the current NAHE recommendations be carried out. Thus, an ongoing control of both the 28-day chloride diffusivity and the concrete cover had to be carried out for all the concrete structures. Therefore, for each concrete structure and each concrete quality, calibration curves relating the chloride diffusivity and the electrical resistivity had to be established before concrete construction started. These calibration curves were based on parallel testing of chloride diffusivity and electrical resistivity after 14, 28, 56, and 90 days of water curing at 20°C in the laboratory, a typical example of which is shown in Figure 9.11. For each concrete quality, some additional concrete specimens were also kept water cured in the laboratory for further testing of the chloride diffusivity for periods of up to one year. For the regular concrete quality control, all measurements of the electrical resistivity were based on the four-electrode method (Wenner).

On the construction site, all control measurements of the achieved chloride diffusivity during the construction period were primarily based on a

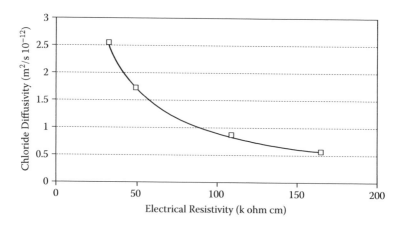

Figure 9.11 A typical calibration curve for control of chloride diffusivity based on electrical resistivity measurements.

Figure 9.12 Production of a wall type of dummy element on the construction site.

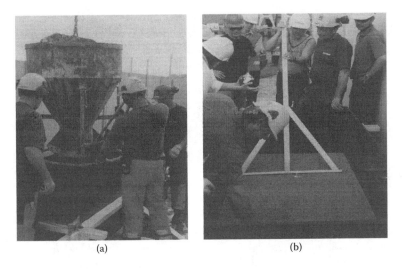

(a)   (b)

Figure 9.13 Production of a slab type of dummy element on the construction site.

number of concrete cores removed from the given structures during concrete construction. In addition, a number of cores from corresponding dummy elements separately produced on the construction site were also tested. For each concrete structure, one or more dummy elements were produced, and these were either a slab or a wall type of element, or both (Figures 9.12 and 9.13). A typical development of chloride diffusivity on the construction site and in the laboratory for up to one year is shown in Figure 9.14.

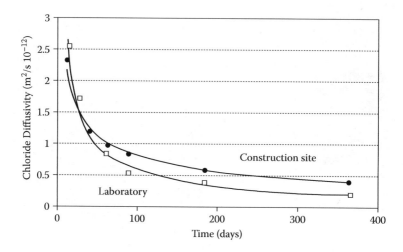

Figure 9.14 Development of chloride diffusivity on the construction site and in the laboratory for up to one year.

For each concrete structure, ongoing control measurements of the achieved concrete cover after concreting were also carried out during concrete construction. All these measurements were manually carried out on protruding steel bars in construction joints.

9.4.4 Achieved construction quality

9.4.4.1 Compliance with specified durability

For all the concrete structures in the first four parts of the project (Contractor A), a probability of corrosion as low as possible and not exceeding 10% for a 150-year service period was specified. To show compliance with such a durability requirement, a new durability analysis had to be carried out upon completion of each structure. These analyses were carried out with input parameters based on the obtained control data for both the 28-day chloride diffusivity and the concrete cover. For these analyses, all the other previously assumed input parameters were kept constant. Hence, this documentation primarily reflected the results obtained from the regular quality control during concrete construction, including the scatter and variability involved. For all the structures where a given value for the 28-day chloride diffusivity had been specified, any unacceptable deviation from this value could be detected and corrected for during concrete construction. The same was true for the regular and ongoing control of the achieved concrete cover.

For the first concrete substructure in Part 1 of the project, a type of concrete was applied that was somewhat retarded compared to that of the intended type of concrete. Therefore, the obtained average 28-day chloride diffusivities of 3.0 and 5.0 × 10⁻¹² m²/s for the bottom slab and the external walls of this structure, respectively, were higher than the intended value of ≤ 2.0 × 10⁻¹² m²/s. Since the type of concrete in question showed a very rapid further reduction of chloride diffusivity, however, this concrete was still accepted.

For all the external walls in the first concrete structure where a nominal concrete cover of 100 mm was specified, an average concrete cover of 102 mm with a standard deviation of 8 mm was obtained. For one section of these walls, however, the quality control revealed a distinct deviation. For this particular section, an average concrete cover of only 74 mm with a standard deviation of 8 mm was observed, and as a consequence, a special protective surface coating was required and later applied for this section. For this particular concrete structure as a whole, however, probabilities for corrosion in bottom slab and external walls of 0.24 and 2.1%, respectively, were obtained (Table 9.8). For the open concrete deck in the second concrete structure of Part 1, where stainless steel was applied, the corrosion probability was 0.13%.

For all the other concrete structures in Parts 2 to 4 of the project, the 28-day chloride diffusivity typically varied from 2.0 to 4.1 × 10⁻¹² m²/s, which, in combination with the achieved concrete cover, gave probabilities of corrosion in the bottom slabs and in the external walls typically varying from 0.01 to 0.92% and from less than 0.001 to 0.02%, respectively. Thus, the results in Table 9.8 demonstrate that for all the concrete substructures in Parts 1 to 4 of the project, the specified durability was achieved with a very good margin.

For all the other concrete substructures in parts 5 to 8 of the project that were only based on prescriptive durability requirements (Contractor B), it was not possible to provide any documentation of compliance with the specified durability. Since a performance-based concrete quality control was also carried out for these structures, however, a documentation of the achieved construction quality in the form of corrosion probabilities after 150 years could be obtained. These durability analyses were also based on the obtained average values and standard deviations of both the 28-day

Table 9.8 Obtained probabilities of corrosion (%) based on regular control measurements of the 28-day chloride diffusivity and concrete cover (Contractor A)

Part of project	Bottom slab	External walls	Open deck
1	0.24	2.1	0.13
2	0.92	0.02	—
3	0.64	0.002	—
4	0.01	<0.001	—

Table 9.9 Obtained probabilities of corrosion (%) based on
regular control measurements of 28-day chloride
diffusivity and concrete cover (Contractor B)

Part of project	Bottom slab	External walls	Open deck
5	15	3	6
6	—	11–13	—
7	14	1.3	—
8	—	—	4.5

chloride diffusivity and the concrete cover from the regular quality control of each structure (Table 9.9). For the concrete structure in Part 5, corrosion probabilities in the bottom slab and external walls of typically 15 and 3% were obtained, respectively, while for the open concrete decks, the obtained probability was about 6%. For the concrete structure in Part 6, no control measurements for the bottom slab were carried out, but for the external walls of this structure, corrosion probabilities varying from 11 to 13% were obtained. For the bottom slab and external walls of the structure in Part 7, probabilities of about 14 and 1.3% were obtained, respectively, while for the open concrete decks in Part 8, the probability was about 4.5%.

The generally higher corrosion probabilities obtained for all the concrete substructures in Parts 5 to 8 of the project (Table 9.9), compared to those of the structures in Parts 1 to 4 (Table 9.8), may be ascribed due to several reasons. For all the concrete structures in Parts 1 to 4, the concrete was based on a blast furnace slag cement with 70% slag (CEM III/B 42.5 LH HS) in combination with silica fume, while all the concrete structures in Parts 5 to 8 were produced with concrete based on fly ash cements in combination with silica fume. For most of these structures, a cement with 30% fly ash (CEM II/B-V 32.5 N) was applied, but partly also a cement with 20% fly ash (CEM II/A-V 42.5 N). It is well known that blast furnace slag cements generally show a very rapid reduction of chloride diffusivity by time, while fly ash cements generally show a slower reduction (Chapter 3). For all the external walls in Parts 2 to 4 of the project, stainless steel was also applied, while the much higher probabilities for the bottom slabs in Table 9.9, compared to those in Table 9.8, primarily reflect the different applied concrete covers of 80 and 100 mm, respectively.

Although the mixture compositions of the various types of concrete applied in Parts 5 to 8 of the project were basically the same, the obtained 28-day chloride diffusivities from one construction site to another were quite different. Thus, for the concrete structures in Parts 5 and 7 of the project, which were produced at one construction site, the 28-day diffusivity typically varied from 6.4 to 8.9 $m^2/s \cdot 10^{-12}$, while for the concrete structure in Part 6, which was produced at another construction site, it typically varied from 12.1 to 16.7 $m^2/s \cdot 10^{-12}$.

9.4.4.2 In situ quality

For documenting the achieved in situ quality during the construction period, a number of concrete cores were tested from both the concrete structures under construction and the corresponding dummy elements. Based on the obtained chloride diffusivities after one year of site curing combined with the obtained control data on concrete cover as input parameters, new durability analyses were carried out for each concrete structure. Also, all the other previously selected input parameters to the analyses were held constant. The typical values of the achieved in situ quality site after one year are shown in Table 9.10.

For all the concrete substructures in Parts 1 to 4 of the project (contractor A), it can be seen from Table 9.10 that very low corrosion probabilities were obtained compared to those of the structures in Parts 5 to 8. For both the bottom slabs and external walls of the concrete structures in Parts 1 to 4, the corrosion probability was typically less than 0.001% and hardly detectable, while for the concrete structures in Parts 5 to 8 (Contractor B), the corrosion probability for the bottom slabs and the external walls typically varied from 20 up to 70% and from 0.6 to 30%, respectively. Also, for the open concrete decks, a high variation in corrosion probability was obtained. The generally slow development of chloride diffusivity for the concrete based on fly ash cements has already been pointed out. In particular, this was true for those concrete structures that were produced during the winter seasons at low curing temperatures. For marine construction work, this may have some implications for early-age exposure of the concrete to seawater before the concrete has gained sufficient maturity and density, as shown in Chapter 2.

For the concrete structure in Part 6, it should also be noted that the in situ data on achieved chloride diffusivity were only based on concrete cores from the separately produced dummy element. Thus, the obtained probability of 30% for the external walls of this structure is not very representative

Table 9.10 Obtained probabilities of corrosion (%) based on in situ data after one year on the construction site

Part of project	Bottom slab	External walls	Open deck
I	<0.001	<0.001	0.02
2	<0.001	<0.001	—
3	<0.001	<0.001	—
4	<0.001	<0.001	—
5	70	25	35
6	—	30	—
7	20	0.6	—
8	—	—	1.2

for this particular concrete structure. For one of the external walls of this structure, however, a severe segregation of the self-consolidating concrete during concrete construction took place. Separate investigations based on extensive concrete coring of this particular wall showed that the in situ strength of the concrete was still acceptable, but the durability properties were distinctly reduced.

9.4.4.3 Potential quality

For most types of binder system, the development of chloride diffusivity tends to plateau after about one year of water curing at 20°C in the laboratory. In order to provide information about the potential construction quality of the various structures, the chloride diffusivity was also determined on a number of separately produced and water-cured specimens in the laboratory for up to one year. These chloride diffusivities combined with the achieved site data on concrete cover were used as new input parameters to further durability analyses. Also, all the other originally assumed input parameters to the durability analyses were kept constant. Typically achieved values for the potential quality of the various concrete structures are shown in Table 9.11.

Both for the bottom slabs and the external walls of all the concrete substructures in Parts 1 to 4 of the project (Contractor A), the corrosion probability was typically less than 0.001% and hardly detectable. For all these structures, therefore, the potential construction quality was extremely good. Also, for the concrete structures in Parts 5 to 8 of the project (Contractor B), the corrosion probability was very low, typically varying from 0.04 to 0.5% and from 0.01 to 0.05% for the the bottom slabs and the external walls, respectively, but distinctly higher than for the concrete structures in Parts 1 to 4 of the project. These results clearly demonstrate

Table 9.11 Obtained probabilities of corrosion (%) based on laboratory-produced specimens water cured in the laboratory for one year

Part of project	Bottom slab	External walls	Open deck
1	<0.001	<0.001	0.002
2	<0.001	<0.001	—
3	<0.001	<0.001	—
4	<0.001	<0.001	—
5	0.04	0.01	0.01
6	—	0.05	—
7	0.5	0.01	—
8	—	—	0.5

that the concrete structures based on high-volume fly ash cements also reached a good potential construction quality as long as good curing conditions were provided.

9.4.5 Frost resistance

Since the frost resistance of concrete based on blast furnace slag cements is generally considered to be lower than that of concrete based on other types of binder systems, an extensive documentation program on the frost resistance of the slag cement type of concrete was carried out (Årskog and Gjørv, 2010).

Typically, the concrete in question had a cement content of 390 kg/m³ (CEM III/B 42.5 LH HS) in combination with 39 kg/m³ silica fume (10%), giving a water/binder ratio of 0.37. In order to improve the fresh concrete properties, a small amount of entrained air (3%) had been added to the concrete. In addition to this basic type of concrete, the documentation program also included the testing of two other versions of the concrete mixture, one with a higher dosage of entrained air (6%) and, as a reference, a concrete without any air entrainment. The documentation program was further based on two different types of test methods with very different exposure conditions, one of which was the German CDF method prEN12390-9 with 28 freeze-thaw cycles (Setzer et al., 1996), while the other was the Swedish method SS 13 72 44-3 with 112 freeze-thaw cycles (SSI, 1995). From both test methods it was concluded that all three versions of the concrete based on the 70% slag cement showed a very good frost resistance regardless of the varying air-void content.

The above test results are in general agreement with other experiences reported in the literature; if only the concrete is made sufficiently dense, e.g., water/binder ratio of 0.40 or less, the frost resistance of concrete based on blast furnace slag cements should not represent any durability problem (Gjørv, 2012). Based on the above documentation, therefore, the given concrete based on 3% entrained air was adopted for all the concrete structures produced in Parts 1 to 4 of the project (Contractor A).

For all the concrete structures produced with 30% fly ash cement (CEM II/B-V 32.5 N) in Parts 5 to 8 of the project, no documentation on the frost resistance was provided. For all the structures produced with both 30 and 20% fly ash cement (Contractor B), the required frost resistance was based on an intended air-void content of 4 to 6%. The regular concrete quality control revealed, however, that it was very difficult to keep the intended air-void content during concrete construction; the air-void content typically showed a high scatter and variability, which is often the case for concrete based on fly ash cements with a varying content of carbon (Gebler and Klieger, 1983; Nagi et al., 2007).

9.4.6 Additional protective measures

For all external walls of the first concrete structure in Part 1 of the project (Contractor A), provisions for future installation of cathodic prevention had been applied as an additional protective measure. Since such a protective measure requires that the ohmic resistance between any two points in the rebar system does not exceed 1 ohm (CEN, 2000), a special quality program was carried out during concrete construction showing that a proper electrical continuity was obtained. For this particular structure, the specification also included embedded probes and instrumentation for future control of chloride penetration, but these probes were only embedded in one selected location of the structure. For all the other concrete substructures in the first four parts of the project (Contractor A), the additional protective measure was based on a partial replacement of the black steel with stainless steel of type W.1.4301.

Since all the prefabricated caissons produced in Parts 5 to 8 of the project (Contractor B) were installed on structural steel tubes and elements that were cathodically protected, these cathodic protection systems were designed in such a way that they also cathodically protected the submerged parts of the installed concrete structures. Also for these concrete structures, the quality control showed a proper electrical continuity within the rebar system. For those parts of the concrete structures that were located above water, however, only provisions for future installation of cathodic prevention in combination with embedded probes for control of chloride ingress in one location for each structure were applied.

9.5 EVALUATION AND DISCUSSION OF OBTAINED RESULTS

For a more complete durability design of concrete structures in marine environments, it should be noted that potential durability problems other than chloride-induced corrosion must also be properly considered and taken into account. The same is true for control of early-age cracking. Based on the above procedures for durability design and quality assurance, however, the specified durability and the results obtained on achieved construction quality for all the above case structures are briefly summarized in Table 9.12. Since the risk for corrosion is primarily related to those parts of the structures that are located above water, the results from the structures in the New City Development in Table 9.12 only include these parts of the structures. For this particular project, Table 9.12 also only shows the results obtained from those structures that were subjected to the probability-based durability design.

Again, it should be noted that the above procedures for durability design do not provide any basis for prediction or assessment of service life of the given structures. Beyond onset of corrosion, a very complex deteriorating

Table 9.12 Specified service periods and achieved construction quality based on corrosion probability

Project	Specified service period (years)	Achieved construction quality (corrosion probability%)		
		Compliance	In Situ Quality	Potential Quality
Container Terminal 1, Oslo (2002)	100 years: Current concrete codes	—	After 100 years: Approx. 80%	After 100 years: Approx. 60%
Container Terminal 2, Oslo (2007)	100 years: Probability of corrosion ≤ 10%	OK; approx. 2%	After 100 years: Approx. 0.05%	After 100 years: <0.001%
New City Development, Oslo (2010)	150 years: Probability of corrosion ≤ 10%	OK; 0.02 < 0.001%[a]; approx. 2%[b]	After 150 years: <0.001%[a]; <0.001%[b]	After 150 years: <0.001%[a]; <0.001%[b]

[a] Stainless steel.
[b] Black steel.

process starts with many further critical stages before the service life is reached. As soon as the first chlorides have reached embedded steel and corrosion starts, however, the owner of the structure has gotten a problem, which at an early stage only represents a maintenance and cost problem, but later may also gradually develop into a more difficult controllable safety problem. As a basis for the durability design, therefore, efforts should be made to obtain a best possible control of the chloride ingress during the initiation period before any corrosion starts. It is in this early stage of the deteriorating process that it is both technically easier and much cheaper to take necessary precautions and select proper protective measures for control of the further deteriorating process. Such control has also shown to be a very good strategy from a sustainability point of view (Chapter 11).

Since the above calculations of corrosion probability are based on a number of assumptions and simplifications, it should be further noted that the obtained service periods with a corrosion probability of less than 10% should not be considered real service periods for the given structures. However, for all the case structures where the above procedures for durability design were applied, the durability analyses supported an engineering judgment of the most important parameters related to the durability, including the scatter and variability involved. Hence, a proper basis for comparing and selecting one of several technical solutions for obtaining a best possible durability of the given concrete structures in the given environments was obtained. As a result, performance-based durability requirements could be specified that could be verified and controlled for quality assurance during concrete construction.

For the reference project (Container Terminal 1), where the specified 100-year service life was only based on descriptive durability requirements according to the current concrete code, it was not possible to provide any documentation of compliance with the specified durability. However, based on the applied concrete, which typically showed a 28-day chloride diffusivity of 6.0×10^{-12} m^2/s and an achieved concrete cover of 65 mm, it was possible to estimate a service period of about 30 years before the probability of corrosion would reach 10%. Also, after a 100-year service period, the probability of corrosion would reach about 65%, while based on the achieved chloride diffusivities on the construction site and in the laboratory during a period of six months, in combination with the site data on concrete cover, respective corrosion probabilities of about 80 and 60% would be reached.

For this particular concrete harbor structure, it should be noted that during concrete construction, heavy winds and high tides also occurred. As a result, a deep chloride ingress in several of the freshly cast deck beams took place already during concrete construction, as shown in Figure 2.28 (Chapter 2). Therefore, if the risk for such an early-age chloride exposure before the concrete has gained sufficient maturity and density is high, special precautions or protective measures should be considered.

Based on general experience, however, the above results on achieved construction quality for Container Terminal 1 indicate that a relatively high level of operation costs would be expected for maintaining a proper long-term performance and serviceability of the given structure during the specified service life of 100 years.

For Container Terminal 2 a 100-year service period with a probability for corrosion of less than 10% was specified. Based on obtained average data for the 28-day chloride diffusivity of 7.9×10^{-12} m^2/s and concrete cover of 99 mm from the quality control, a probability for corrosion of about 2% after a 100-year service period was achieved. Thus, the specified durability was obtained with a proper margin. Based on the obtained chloride diffusivity on the construction site and in the laboratory during one year, corrosion probabilities after a 100-year service period of about 0.05 and less than 0.001%, respectively, were achieved. These results indicate that both the achieved in situ quality during the construction period and the potential quality of the structure were very good.

For all the concrete structures in the New City Development on Tjuvholmen, for which a probability of corrosion as low as possible and not exceeding 10% during a 150-year service period was specified, it can be seen from Table 9.12 that this specification was obtained with a very good margin (Contractor A). Thus, for the first concrete structure, which was only produced with black steel reinforcement, a probability of about 2% after a 150-year period was obtained, while for all the other concrete structures, which were produced with a partial replacement of the black steel with stainless steel reinforcement, the corrosion probability typically varied from 0.02 to less than 0.001%. Based on the chloride diffusivities obtained both on the construction site and in the laboratory during one year, the probability of corrosion after a 150-year period was typically less than 0.001% and hardly detectable. These results indicate that both the achieved in situ quality during the construction period and the potential quality of the structures were extremely good.

For all the concrete structures in the New City Development project where the specified durability was only based on prescriptive durability requirements, it was not possible to provide any documentation of compliance with the specified durability (Contractor B). Also, the achieved construction quality for these structures typically showed a higher scatter and variability than all the concrete structures in the first four parts of the project. Based on the achieved control data on both the 28-day chloride diffusivity and the concrete cover, corrosion probabilities after 150 years of 1.3 to 13% were obtained. Based on the achieved chloride diffusivities on the construction site and in the laboratory during one year, corrosion probabilities typically varying from 0.6 to 70% and from 0.01 to 0.05% after 150 years were obtained, respectively. These results indicate that the potential durability of the structures was quite good, but the achieved in situ quality after one year on the construction site was partly very poor. When the prefabricated concrete caissons were pulled out from the dry dock and exposed to seawater at an early stage, this may have had some implications for an early-age exposure to chlorides before the concrete had gained sufficient maturity and density.

For the above concrete structures in the New City Development project where the durability was based on prescriptive durability requirements, the mixture composition of all the applied types of concrete was basically the same, but the achieved concrete quality typically varied from one construction site to another. When severe segregation in one of the concrete structures also took place during concrete construction, the owner of the project had to accept the reduced durability properties of this concrete as long as the code requirement for the in situ compressive strength was still fulfilled. It was very difficult to argue against a contract based on prescriptive durability specifications that could not be verified and controlled during concrete construction.

In order to ensure the required long-term performance of all the concrete structures in the New City Development project, some additional protective measures were also applied. For some of the structures, the protective measure was based on partial use of stainless steel, while for others, cathodic prevention of the submerged parts of the structures in combination with provisions for future cathodic prevention above water was applied. The provisions for cathodic prevention were also combined with embedded instrumentation for future control of the chloride ingress.

For all concrete structures in marine environments, extensive experience demonstrates that the risk for corrosion is primarily related to those parts of the structures that are located above water, and if these parts of the structures are cathodically protected right from the beginning, a cathodic prevention has shown to be quite effective (Chapter 5). If only provisions for cathodic prevention are applied, however, both the installation and the activation of such a preventive system must be implemented before the first chlorides have reached the embedded steel and corrosion starts. As a consequence, a very close control of the chloride ingress during operation of the structures must be carried out, which may represent a great challenge to the owner. Also, for all the concrete structures in the current project, the achieved construction quality typically showed a high scatter and variability, while all the probes for future control of the chloride ingress were only embedded in one particular location of each structure.

For all the structures where a partial use of stainless steel was applied, this protective measure showed to be a very simple and robust technical solution. For all these concrete structures, stainless steel in the most exposed and vulnerable parts of the structures also proved to be economically competitive with the provisions for future cathodic prevention in combination with embedded instrumentation, even on a short-term basis. On a long-term basis, additional high expenses for both installation and operation of a cathodic prevention system are involved (Chapter 5).

9.6 CONCLUDING REMARKS

As clearly demonstrated in Chapter 2, the durability of concrete structures in severe environments is related not only to design and materials, but also to construction. Many durability problems that develop after some time can be attributed to an absence of proper quality control and special problems during concrete construction. Therefore, the achieved construction quality and variability must be firmly grasped before an increased and more controlled durability and service life of important concrete infrastructure can be reached.

For all the case structures where the above probability-based durability design and concrete quality control were carried out, the specified durability was achieved with a proper margin. For the owners of the structures, it was very important to receive this documentation of compliance before the structures were formally handed over from the contractors. The required documentation of achieved construction quality also clarified the responsibility of the contractors for the quality of the construction process; the required documentation of achieved construction quality clearly resulted in improved workmanship with less scatter and variability of achieved construction quality.

Even if the strongest durability requirements are both specified and achieved during concrete construction, extensive experience demonstrates that for all concrete structures in chloride-containing environments, a certain rate of chloride ingress will always take place during operation of the structures. Upon completion of each structure, therefore, it was also very important for the owners to receive a service manual for future condition assessment and preventive maintenance of the structures. Such a service manual helps provide the ultimate basis for achieving a more controlled durability and service life of the given concrete structure in the given environment.

REFERENCES

Årskog, V., and Gjørv, O.E. (2010). Slag Cements and Frost Resistance. In *Proceedings, Sixth International Conference on Concrete under Severe Conditions—Environment and Loading*, vol. 2, ed. P. Castro-Borges, E.I. Moreno, K. Sakai, O.E. Gjørv, and N. Banthia. Taylor & Francis, London, pp. 795–800.

CEN. (2000). *EN 12696: Cathodic Protection of Steel in Concrete*. European Standard CEN, Brussels.

Gebler, S., and Klieger, P. (1983). *Effect of Fly Ash on the Air-Void Stability of Concrete, Fly Ash, Silica Fume, Slag and Other Mineral By-Products in Concrete*, ACI SP-79, ed. V.M. Malhotra, pp. 103–142.

Gjørv, O.E. (2012). *Blast Furnace Slag for Durable Concrete Infrastructure in Marine Environments, Workshop Proceeding 10, Durability Aspects of Fly Ash and Slag in Concrete*. Nordic Concrete Federation, Tekna, Oslo, pp. 67–81.

Nagi, M.A., Okamoto, P.A., Kozikowski, R.L., and Hover, K. (2007). *Evaluating Air-Entraining Admixtures for Highway Concrete*, NCHRP Report 578. Transportation Research Board, Washington, DC.

NAHE. (2004a). *Durable Concrete Structures—Part 1: Recommended Specifications for New Concrete Harbour Structures*. Norwegian Association for Harbour Engineers, TEKNA, Oslo (in Norwegian).

NAHE. (2004b). *Durable Concrete Structures—Part 2: Practical Guidelines for Durability Design and Concrete Quality Assurance*. Norwegian Association for Harbour Engineers, TEKNA, Oslo (in Norwegian).

NAHE. (2004c). *Durable Concrete Structures—Part 3: DURACON Software*. Norwegian Association for Harbour Engineers, TEKNA, Oslo.

NORDTEST. (1999). *NT Build 492: Concrete, Mortar and Cement Based Repair Materials: Chloride Migration Coefficient from Non-Steady State Migration Experiments*. NORDTEST, Espoo, Finland.

NPRA. (1996). *Handbook 185*. Norwegian Public Roads Administration—NPRA, Oslo (in Norwegian).

Setzer, M.J., Fagerlund, G., and Janssen, D.J. (1996). CDF Test—Test Method for the Freeze-Thaw Resistance of Concrete—Tests with Sodium Chloride Solution (CDF). *Materials and Structures*, 29, 523–528.

SSI. (1995). *SS 13 72 44-3: Concrete Testing—Hardened Concrete—Scaling at Freezing*. Swedish Standards Institution, Stockholm (in Swedish).

Standard Norway. (2003a). *NS-EN 206-1: Concrete Part 1. Specification, Performance, Production and Conformity*, Amendment prA1:2003 Incorporated. Standard Norway, Oslo (in Norwegian).

Standard Norway. (2003b). NS 3473: *Concrete Structures—Design and Detailing Rules*. Standard Norway, Oslo (in Norwegian).

Standard Norway. (2003c). *NS 3465: Execution of Concrete Structures—Common Rules*. Standard Norway, Oslo (in Norwegian).

Chapter 10

Life cycle costs

10.1 GENERAL

Calculations or estimations of costs against benefits can be carried out in different ways by considering various types of cost or benefit, and this is often referred to in terms of whole life cycle costing, cost-benefit analysis, or cost-benefit-risk analysis. Life cycle costs (LCCs) may be used as a valuable tool for assessment of the cost-effectiveness of various technical solutions for an optimal durability design. It may also be used for assessment of various technical solutions for condition assessment, maintenance, and repair strategies during operation of the structure.

As a basis for the life cycle costs of a concrete structure up to the time t_N, the following expression may be used:

$$LCC(t_N) = C_I + C_{QA} + \tag{10.1}$$

$$\sum_{i=1}^{t_N} \frac{C_{IN}(t_i) + C_M(t_i) + C_R(t_i) + \sum_{LS=1}^{M} p_{f_{LS}}(t_i) \cdot C_{f_{LS}}}{(1+r)^{t_i}}$$

where C_I is the design and construction cost, C_{QA} is the cost of quality assurance and quality control, $C_{IN}(t)$ is the expected cost of inspections, $C_M(t)$ is the expected maintenance cost, $C_R(t)$ is the expected repair cost, M is the number of limit states (LS), $p_{f_{LS}}(t)$ is the annual probability of failure for each limit state, $C_{f_{LS}}$ is the failure cost associated with the occurrence of each limit state, and r is the discount rate.

However, the above calculation of life cycle costs does not account for the advantage of designing or maintaining the structure for achieving a longer service life. Therefore, it may be more meaningful to compare all costs on an annual-equivalent basis by distributing all life cycle costs over the whole lifetime of the structure. This can be done by use of an annuity factor that expresses the annuity or annual costs. The average annuity cost (C_A) during the service life of a structure (n years) can then be expressed as

$$C_A(t_N) = \sum_{j=1}^{n} \frac{p_f(t_j) \cdot r \cdot \left[C_I + C_{QA} + C_{IN}(t_j) + C_M(t_j) + C_R(t_j) \right]}{1 - (1+r)^{-t_j}} \quad (10.2)$$

where $p_f(t_j)$ represents the probability of failure in year (j) and

$$p_f(t_n) = 1 - \sum_{j=1}^{n-1} p_f(t_j) \quad (10.3)$$

Experience has shown that calculations of annuity costs represent a more appropriate way of expressing increased investment costs for increased durability. Calculations of life cycle costs may also include other costs or benefits, such as traffic delays or reduced travel time, as well as efficiency of inspections, maintenance and repair strategies, etc. Evidently, the decision analysis should be the subject for a sensitivity analysis in order to ensure that decisions are not unduly influenced by the uncertainties in the various types of costs.

10.2 CASE STUDY

10.2.1 General

In order to demonstrate how an assessment of life cycle costs may provide a further basis for decision making of various technical solutions for improved durability, a heavily corroding concrete structure in Trondheim Harbor was selected for a case study. The structure, which was constructed in 1964, was a traditional concrete harbor structure with an open concrete deck of 132 × 17 m on top of slender underwater-cast concrete pillars. The concrete deck has 3 longitudinal main girders and 18 transversal secondary beams with two-way slabs in between. The main girders and the secondary beams have dimensions of 90 × 120 cm and 70 × 70 cm, respectively, while the top slab is 25 cm thick, including a 6 cm top layer. The mechanical loads on the deck consist mainly of two heavy loading cranes with capacities of 60 and 100 tons, respectively, moving on top of the three longitudinal main girders (Figure 10.1).

After a service period of about 38 years (2002), the general condition of the structure was very poor due to heavy corrosion of the embedded steel. A structural assessment confirmed that the load-bearing capacity of the main girders would only be acceptable for continued operation of the heavy

Figure 10.1 A heavily corroding concrete harbor structure from 1964 in Trondheim. (Courtesy of Trondheim Harbor KS.)

loading cranes for a very short period of time. The observed rate of corrosion in the girders was so high that an immediate repair was considered very urgent. The concrete pillars were in fairly good condition, but the deck slabs had already reached such a high degree of deterioration that all other traffic on the concrete deck had been prohibited for some time. Of the various technical solutions for repair considered, one option would simply be to construct a new concrete deck on top of the old deck, only using the old deck as a formwork. However, since both the heavy crane facilities and the harbor structure as a whole would not be needed for a continued operation of more than a very limited period of time, the construction of a new concrete deck would represent a too expensive solution. Therefore, a specially designed cathodic protection system was applied in order to reduce the rate of corrosion in the three main girders as much as possible, and thus extend the service life of the structure for a limited period of time (Vælitalo et al., 2004).

If the given concrete harbor structure originally had been the subject for a proper durability design, including assessment of life cycle costs, a more controlled durability and service life could have been obtained. If the objective for such a durability design had been to keep a safe operation of the structure for a service period of approximately 50 years, the following life cycle costs of various technical solutions could have been considered. In order to demonstrate the usefulness of such calculations, a very simple cost comparison of various possible technical solutions was carried out. In order to provide a basis for such calculations, some technical data about the old structure were needed. Although such data were not easily available, some information was provided as shown in Table 10.1.

Table 10.1 Basic information about the existing concrete structure

Concrete quality	45 MPa
Concrete cover in beams and girders	75 mm
Concrete cover in slabs	25 mm
Assumed new construction cost	25,000,000 NOK
Amount of concrete	1532 m³
New cost of concrete (1.200 NOK/m³)	1,840,000 NOK
Amount of steel	315 tons
New cost of steel (3.650 NOK/t)	1,150,000 NOK
Material costs related to total costs:	
Concrete	7.4%
Steel	4.6%

In order to demonstrate the principles for the cost calculations, the following alternative options were considered:

- Doing nothing other than what was originally designed
- Increased concrete quality from 45 to 70 MPa
- Increased nominal concrete cover from 75 to 100 mm in beams and girders
- Increased concrete quality from 45 to 70 MPa in combination with increased nominal concrete cover from 75 to 100 mm
- Partial use of stainless steel as reinforcement in beams and girders (75%)
- Use of stainless steel as reinforcement in beams and girders (100%)

In the following, the life cycle costs of all the above options were compared for a service period of 50 years. For convenience, the discount rate was put to zero in all calculations. Annuity costs were therefore calculated by the total costs divided by the expected service life. Other maintenance costs that usually come along during operation of such a structure were not included.

10.2.2 Doing nothing

The total life cycle cost for this option was NOK 25,000,000. It was assumed that the service life of the existing structure would end after a continued service period of approximately three years (2005), which means that the annuity cost was calculated to be approximately NOK 630,000.

10.2.3 Increased concrete quality

By use of an increased concrete quality from 45 to 70 MPa based on a pure portland cement, a durability analysis indicated an increased service life of the existing structure by up to 10 years. The material cost for such increased quality of concrete would be approximately NOK 2,200,000. Hence, an increased cost of NOK 380,000 would increase the total cost by 1.5% to NOK 25,380,000. For an assumed extended service life of approximately 10 years, the annuity cost was calculated to be approximately NOK 500,000.

10.2.4 Increased concrete cover

Based on a concrete quality of 45 MPa, a durability analysis indicated that an increased nominal concrete cover of up to 100 mm in beams and girders would be needed in order to reach a service life of approximately 50 years. By an increased nominal concrete cover from 75 to 100 mm, the additional material cost for this increased concrete cover would be approximately NOK 70,200 (58.5 m³ concrete), which would give an increased total cost by 0.2% to NOK 25,070,200. For an extended service life by approximately 10 years, the annuity cost was calculated to approximately NOK 500,000.

10.2.5 Increased concrete quality and concrete cover

By combining the beneficial effects of increased concrete quality from 45 to 70 MPa and increased nominal concrete cover from 75 to 100 mm in all beams and girders, an estimated extended service life of approximately 25 years would have been achieved. The material cost for such a solution would be approximately NOK 2,420,000, giving an additional cost of NOK 580,000. This increased cost would increase the total cost by 2.3% to NOK 25,580,000. For an assumed extended service life of approximately 25 years, the annuity cost was calculated to be approximately NOK 380,000.

10.2.6 Seventy-five percent stainless steel reinforcement

As shown in Chapter 2 and discussed in Chapter 5, stainless steel reinforcement has proved to perform extremely well in marine environments for a very long period of time. In the literature, it is generally assumed that proper use of stainless steel will increase the service life by a factor of at least two (Cramer et al., 2002). A technical option could therefore have been to select stainless steel reinforcement. By using 75% of stainless steel in all beams and girders with a cost ratio to black steel of 4.5 (Cramer et al., 2002), the material cost for the reinforcement would be approximately NOK 4,170,000, which would increase the total cost by 12.1% to

NOK 28,020,000. For an assumed extended service life of at least 40 years, the annuity cost was calculated to less than NOK 350,000.

10.2.7 One hundred percent stainless steel reinforcement

By using 100% stainless steel in all beams and girders, the material cost for the reinforcement would increase to approximately NOK 5,200,000, which again would increase the total cost by 16.1% to NOK 30.200.000. With an assumed extended service life of at least 40 years, the annuity cost was calculated to less than NOK 380,000.

10.2.8 Cathodic protection

After 38 years of service, a specially designed cathodic protection system was developed and installed for the given structure in order to extend the service life of the three main girders by approximately 15 years (Vælitalo et al., 2004). Since the cost of this installation was approximately NOK 3,000,000, the annuity cost for this situation was calculated to approximately NOK 540,000.

10.2.9 Evaluation and discussion of obtained results

A summary of the results from the above cost calculations is shown in Table 10.2, from which it can be seen that a proper utilization of stainless steel in all beams and girders would have given distinctly lower annuity costs than those of all the other technical solutions.

The results from the above case study demonstrate that assessment of life cycle costs is a very valuable tool for improved decision making in the

Table 10.2 Comparison of life cycle costs for various technical solutions for ensuring a proper operation of the concrete harbor structure during a service period of 50 years

	Additional service life (years)	$LCC(t_{end})$ (%)	Annuity costs $C_A(t_n)$ ($\times 10^6$ NOK)
Doing nothing	0	100.0	0.63
Increased concrete quality	+10	101.5	0.50
Increased concrete cover	+10	100.2	0.50
Increased concrete quality and concrete cover	+25	102.3	0.38
75% stainless steel rebars	>40	112.1	<0.35
100% stainless steel rebars	>40	116.1	<0.38
Cathodic protection	+15	112.0	<0.54

durability design. Even for a service period of only 50 years, a proper utilization of stainless steel would have given a very safe and cost-effective operation compared to that of the traditional design with the corresponding repair and maintenance costs. As already discussed in Chapter 5, such a simple and robust technical solution would ensure proper durability.

REFERENCES

Cramer, S.D., Covino Jr., B.S., Bullard, S.J., Holcomb, G.R., Russell, J.H., Nelson, F.J., Laylor, H.M., and Soltesz, S.M. (2002). Corrosion Prevention and Remediation Strategies for Reinforced Concrete Coastal Bridges. *Cement and Concrete Composites*, 24, 101–117.

Vælitalo, S.H., Pruckner, F., Ødegård, O., and Gjørv, O.E. (2004). A New Approach to Cathodic Protection of Corroding Concrete Harbor Structures. In *Proceedings, Fourth International Conference on Concrete under Severe Conditions— Environment and Loading*, vol. 1, ed. B.H. Oh, K. Sakai, O.E. Gjørv, and N. Banthia. Seoul National University and Korea Concrete Institute, Seoul, pp. 1873–1880.

Chapter 11

Life cycle assessment

11.1 GENERAL

In recent years, there has been a rapidly increasing concern on how human activities affect the environment by loss of the biodiversity and thinning of the stratospheric ozone, as well as climate changes and consumption of natural resources. The term *sustainable development* was introduced in the final report of the Brundtland Commission (World Commission on Environment and Development (WCED)) in 1987, where *sustainable development* was defined as "development that meets the needs of the present without comprising the ability of future generations to meet their own needs."

On the basis of weight, volume, and money, the construction industry is the largest consumer of materials in our society. Thus, approximately 40% of all materials used are related to the construction industry (Ho et al., 2000). From a production point at view, several of the construction materials have a great impact on both the local and the global environment. This is particularly true for concrete, as one of the most dominating construction materials. Therefore, an increased environmental consciousness in the form of better utilization of concrete as a construction material and the creation of a better harmony and balance with our natural environment represents a great and increasing challenge to the construction industry. This was the conclusion from two international workshops that focused on this problem in 1996 and 1998, respectively, the first of which resulted in the Hakodate Declaration (Sakai, 1996), and later on, the Lofoten Declaration (Gjørv and Sakai, 2000), which stated:

We concrete experts shall direct concrete technology towards a more sustainable development in the 21st century by developing and introducing into practice:

(1) Integrated performance-oriented life cycle design
(2) More environmental-friendly concrete construction
(3) Systems for maintenance, repair and reuse of concrete structures. In addition, we shall share information on all these issues with technical groups and the general public.

Table 11.1 Typical energy consumption and gas emission for production of portland cements

Energy consumption (MJ/kg)	Electricity	0.56
	Fossil fuel	3.48
Total		4.05
Emission to the atmosphere (g/kg)	CO_2	867
	SO_2	0.29
	NO_x	1.98
	VOC	0.02
	Dust	0.17

Source: Gjørv, O. E., Durability Requirements in Norwegian Concrete Codes in a Resource and Environmental Perspective, in *Proceedings, Annual Conference of the Norwegian Concrete Association,* Oslo, Norway, 1999 (in Norwegian).

In addition to the large consumption of natural resources for concrete production, the production of pure portland cements is based on a very energy consuming and polluting industrial process. Although distinct improvements in the production technology of cements have been made in recent years, the production of each ton of portland cement still typically requires an energy amount of approximately 4 GJ, while almost 1 ton of CO_2 gas is released to the atmosphere. In addition, a number of other harmful constituents are also released (Table 11.1). On a worldwide basis, cement production makes up about 1.4 billion tons of CO_2 every year, which is about 7% of the total global production of CO_2 (Malhotra, 1999). About half of the CO_2 emission from cement production is due to the de-carbonizing of limestone, while the other half is due to the combustion of fossil fuels. Therefore, proper durability design and concrete quality control, as well as preventive maintenance for ensuring an increased and more controlled durability and service life of concrete structures, are of greatest importance from an environmental point of view.

As already discussed in Chapter 2, an uncontrolled and premature deterioration of concrete structures has emerged to be one of the most demanding challenges facing the construction industry. Public agencies are spending significant and rapidly increasing proportions of their construction budgets for repairs and maintenance of existing concrete infrastructure. Repair projects will undoubtedly be subject to increasing economic constraints, so there will be a parallel increase in the consideration of durability during the design and construction phases for new concrete infrastructure. Enhanced durability and service life of new concrete infrastructure not only are important from an economical point of view, but also directly affect the sustainability.

These and other factors have spurred the rapid development of a life cycle assessment (LCA) of the structures. The framework and methodology for quantifying the economic and ecological effects and impacts from design, production, and maintenance of structures are available through current standards such as ISO 14040 (ISO, 2006a) and ISO 14044 (ISO, 2006b). As these standards show, LCA includes assessment of consumption of materials and energy, generation of waste, and emission of pollutants, as well as the accompanying environmental and health risks. Therefore, LCA provides a valuable tool for both quantifying and comparing the effects of various technical solutions for improvements in the design, construction, and operation of new concrete infrastructure.

In the following, a brief outline of the LCA methodology is given. A case study is also included showing how an LCA can be applied to evaluate the environmental impacts of two different maintenance strategies for a given concrete structure.

II.2 FRAMEWORK FOR LIFE CYCLE ASSESSMENT

According to ISO 14040, a life cycle assessment (LCA) is aimed at understanding and evaluating the magnitude and significance of the potential environmental impacts of a product system. A key part of an LCA is the development of a life cycle inventory (LCI), which quantifies emissions and extractions required to produce and use the product. Each item listed in the inventory is then connected to environmental or health damages via impact pathways (Figure 11.1). The many impact pathways include well-known effects such as climate change, stratospheric ozone depletion, photochemical oxidant formation, and acidification, but many other effects are also included. Category endpoints include human health, the natural environment, and resource availability. A category indicator, representing the amount of impact potential, can be located at any place between the life cycle inventory results and the category endpoint.

Conducting an LCA is a very difficult process because the relationship between the external environment and a category endpoint can be very complex, and an endpoint can be affected by multiple emission sources and types. Figure 11.2, for example, shows some (but not all) of the emissions that are known to affect acidification, but those same emissions can also affect eutrophication, ecotoxicity, or other categories. Normally, the LCA will stop at the step before the category endpoint (Figure 11.1), showing only the impact categories that are fairly easy to evaluate. It will then interpret the results from the various category indicators. Further information about the methodological framework for the assessment of environmental impacts is available from the standards ISO 14040 and ISO 14044.

If the product system is a structure, an LCA determines the environmental impacts of human activities throughout the complete life cycle, from extraction of raw materials, through the service life, and ending with

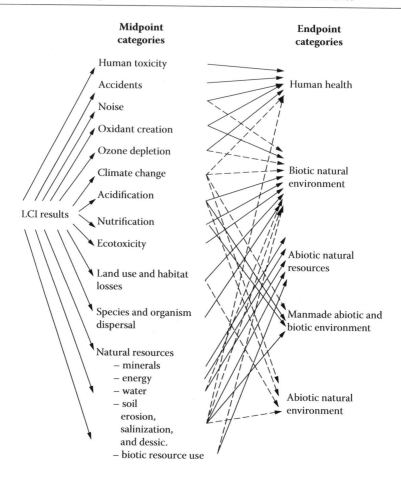

Figure 11.1 General structure of the life cycle inventory analysis framework. Dashed lines indicate that current information for defining the endpoint is particularly uncertain. (From Jolliet, O. et al., *Life Cycle Impact Assessment Programme of the Life Cycle Initiative: Final Report of the LCIA Definition Study*, UNEP/SETSC Life Cycle Initiative, 2003.)

demolition and disposal. Specifically applied to repairs and maintenance of concrete structures, all steps in the repair and maintenance process must be thoroughly evaluated.

The type of repair or maintenance action will be based on a condition assessment and investigation of the given structure. The selected action will depend on the condition of the structure and the external environment, as well as the type of equipment and materials to be used during the process. The next step will be determining the functional unit, which is the reference unit used in the life cycle study (ISO, 2006a, 2006b) All emissions,

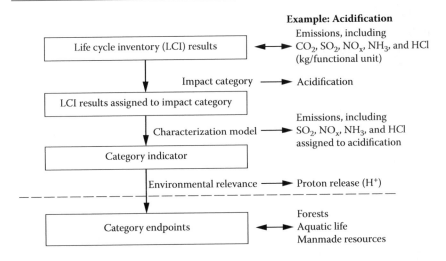

Figure 11.2 Concept of category indicators applied to acidification. (From ISO, *ISO 14044—Environmental Management, Life Cycle Assessment, Requirements and Guidelines*, International Organization for Standardization, Geneva, Switzerland, 2006.)

energy consumption, and flow of materials occurring during the process are related to this unit. The functional unit has to be measurable, and its selection will be affected by the goal and scope of the analysis. The goal of the life cycle assessment must unambiguously state the intended application and indicate to whom the results will be addressed. Thus, the functional unit for a protective surface coating may be defined as the unit of concrete surface that needs to be protected for a specified period of time.

The LCI phase will then consist of the following:

- Quantifying the raw materials, chemicals, and equipment necessary to fulfill the repair or maintenance function. This quantification gives the reference flow (ISO, 2006b), for which all inputs and outputs are referred to and are closely connected to the functional unit.
- Environmental data for the consumed raw materials, chemicals, and equipment from suppliers (specific data), databases (generic data), or an LCI carried out at the supplier level. Materials used should have an environmental declaration with a cradle-to-grave type of scope. The environmental declaration must include use of resources such as energy (renewable or nonrenewable), materials (renewable or nonrenewable), water, waste, and emissions to both air and water.
- Quantifying and classifying the waste from the process, such as recycling or disposal (hazardous or not hazardous).

The calculations should be carried out further by assigning LCI results to impact categories (Figure 11.1). Classification and characterization should be carried out according to ISO 14044, using effect factors from the Montreal Protocol (UN, 1987), IPCC 1995 (UN, 1995) and Heijungs et al. (1992). Emission of a specific gas may be assigned to more than one category. For example, NO_x emissions will be assigned to the categories of both eutrophication and acidification.

The final result may be displayed as impact categories or weighted to an environmental index, where the weighting is the process of converting indicator results of different impact categories by using numerical factors based on value choice. This is an optional element in ISO 14044. Thus, factors from value choices may be based on political targets according to the Kyoto Protocol (UN, 1997) or other similar preferences. Interpretation of the results based on ISO 14044 must identify, qualify, evaluate, and present the findings of significant issues.

11.3 CASE STUDY

11.3.1 General

In order to demonstrate how the above methodological framework can be applied to evaluate the environmental impacts of different maintenance strategies for a given concrete structure, two examples of commonly used systems were analyzed (Årskog et al., 2003), the results of which are briefly outlined and discussed below.

One system is a patch repair based on shotcrete applied to a concrete structure with damage due to ongoing chloride-induced corrosion. Although such a patch repair is not the most efficient way to get ongoing corrosion under control, as previously discussed in Chapter 8, such repairs are still used for such damage (Chapter 2). The other system is a hydrophobic surface treatment, which is a commonly used protective measure for preventive maintenance (Chapter 5).

For both systems, the following common assumptions for the calculation of the ecological impacts were made:

- Transport distance from supplier to the job site of 60 km
- Materials and equipment transported by truck
- Truck diesel fuel consumption of 0.2 kg per ton-km
- A functional unit comprising 1 m^2 of concrete surface area repaired or protected for a period of 10 years

11.3.2 Patch repair

This analysis was based on the following assumptions:

Table 11.2 Energy consumption and ecological impacts of the patch repair

Process	Use of energy MJ/m^2	Global warming kg CO_2 eq/m^2	Acidification g SO_2 eq/m^2	Eutrophication g PO_4 eq/m^2	Photo-oxidant formation g Ethene eq/m^2
			Impact category		
Hydro jetting	677	84	75	1330	266
Cleaning of reinforcement	296	22	4	350	70
Protective coating on reinforcement	35	1.4	19	2.4	3
Application of shotcrete	59	4.4	19	70	14
Transportation	127	10	8	150	30
Total	1194	122	125	1902	383

- Repaired surface area of 30 m²
- Shotcrete rebound of 25%
- Power supply on the construction site based on diesel engines

The various steps of the process included the following:

- Removal of concrete cover to an average depth of 50 mm by high-pressure hydro jetting (at 1000 bar)
- Cleaning of the reinforcing bars by sand blasting
- Applying a protective coating on the reinforcement
- Application of the shotcrete layer
- Applying curing measures to the shotcrete

The consumption of energy and the ecological impacts of the patch repair are summarized in Table 11.2.

11.3.3 Hydrophobic surface treatment

This analysis was based on the following assumptions:

- A hydrophobic agent comprising an iso-octyltriethoxy type of silane in combination with a mineral thickener
- Treated surface area of 150 m²

The various steps of the process included the following:

Table 11.3 Energy consumption and ecological impacts of the hydrophobic surface protection

Process	Use of energy MJ/m²	Global warming kg CO₂ eq/m²	Acidification g SO₂ eq/m²	Eutrophication g PO₄ eq/m²	Photo-oxidant formation g ethene eq/m²
		Impact category			
Production of hydrophobic agent	47	0.295	0.5	6	2
Surface preparation	17	0.013	0.4	7	1
Transportation and surface treatment	12	0.080	0.1	2	66
Long-term degradation		2.171			1
Total	76	2.559	1	15	70

- Preparation of the concrete surface by high-pressure sand blasting (160 bar).
- Application of the hydrophobic agent using a high-pressure sprayer to a thickness of 0.25 mm. It was assumed that only 45% of the hydrophobic agent was applied to the concrete surface, which is equivalent to approximately 500 g/m², while the rest of the agent (approximately 600 g/m²) was emitted to the air (iso-octyltriethoxy silane is volatile and releases ethanol to the atmosphere).

The consumption of energy and the ecological impacts of the hydrophobic surface protection are summarized in Table 11.3.

11.3.4 Evaluation and discussion of obtained results

In order to carry out the above comparison of the ecological impacts caused by the repair and the application of a protective measure, a number of assumptions had to be made. The results indicate, however, that a hydrophobic surface treatment could be repeated more than five times before the ecological impact in the form of photo-oxidant formation would approach that of the patch repair by shotcreting (Table 11.4).

Based on the methodological framework briefly outlined above, LCA appears to be a good but complex tool for the environmental assessment of the design, production, and maintenance of concrete infrastructures. A more complete assessment of all impacts on the environment caused by human activities throughout the whole life cycle of a concrete structure, however, would be even more complex and difficult. However, efforts to

Table 11.4 Comparison of the ecological impacts caused by patch repair
and the protective measure based on hydrophobic surface treatment

	Impact category				
Method	Use of energy MJ/m²	Global warming kg CO_2 eq/m²	Acidification g SO_2 eq/m²	Eutrophication g SO_2 eq/m²	Photo-oxidant formation g ethene eq/m²
Patch repair	1194	122	125	1902	383
Hydrophobic surface treatment	76	2.6	1	15	70

develop more practical applications of LCA have been going on for some time (Sakai, 2005; Aïtcin, 2011; Sakai and Buffenbarger, 2012; Sakai and Noguchi, 2012), and recently, a new International Organization for Standardization (ISO) standard has been introduced (ISO, 2012).

Although the largest environmental gains would be obtained by a more proper durability design and construction of new concrete infrastructures, the above case study demonstrates that LCA also provides a good basis for selecting proper strategies for the maintenance of important concrete infrastructures. From a sustainability point of view, it would appear that a strategy of conducting regular condition assessments and preventive maintenance is superior to allowing the structure to reach a stage where repairs are needed.

REFERENCES

Aïtcin, P.C. (2011). *Sustainability of Concrete*. CRC Press, London.

Årskog, V., Fossdal, S., and Gjørv, O.E. (2003). *Quantitative and Classified Information of RAMS and LCE for Different Categories of Repair Materials and Systems Subjected to Classified Environmental Exposures*, Document D5.2 of the Research Project LIFECON—Life Cycle Management of Concrete Infrastructures for Improved Sustainability. European Union—Competitive and Sustainable Growth Programme, Project G1RD-CT-2000-00378.

Gjørv, O.E. (1999). Durability Requirements in Norwegian Concrete Codes in a Resource and Environmental Perspective. In *Proceedings, Annual Conference of the Norwegian Concrete Association*, Oslo (in Norwegian).

Gjørv, O.E., and Sakai, K. (eds.). (2000). *Concrete Technology for a Sustainable Development in the 21st Century*. E & FN Spon, London.

Heijungs, R., Guinée, J.B., Huppes, G., Lankreijer, R.M., Udo de Haes, H.A., Wegener Sleeswijk, A., Ansems, A.M.M., Eggels, P.G., Duin, R. van, and Goede, H.P. de. (1992). *Environmental Life Cycle Assessment of Products: Guide and Backgrounds (Part 1)*. Institute of Environmental Sciences, Centrum voor Milieukunde Leiden, University of Leiden, Netherlands.

Ho, D.W.S., Mak, S.L., and Sagoe-Crentsil, K.K. (2000). Clean Concrete Construction: An Australian Perspective. In *Concrete Technology for a Sustainable Development in the 21st Century*, ed. O.E. Gjørv and K. Sakai. E & FN Spon, London, pp. 236–245.

ISO. (2006a). *ISO 14040—Environmental Management, Life Cycle Assessment, Principles and Framework.* International Organization for Standardization, Geneva, Switzerland.

ISO. (2006b). *ISO 14044—Environmental Management, Life Cycle Assessment, Requirements and Guidelines.* International Organization for Standardization, Geneva, Switzerland.

ISO. (2012). *ISO 13315-1—Environmental Management for Concrete and Concrete Structures—Part 1: General Principles.* International Organization for Standardization, Geneva, Switzerland.

Jolliet, O., Brent, A., Goedkoop, M., Itsubo, N., Mueller-Wenk, R., Peña, C., Schenk, R., Stewart, M., and Weidema, B. (2003). *Life Cycle Impact Assessment Programme of the Life Cycle Initiative: Final Report of the LCIA Definition Study.* UNEP/SETSC Life Cycle Initiative.

Malhotra, V.N. (1999). Making Concrete Greener with Fly Ash. *Concrete International*, 13, 61–66.

Sakai, K. (ed.). (1996). *Integrated Design and Environmental Issues in Concrete Technology.* E & FN Spon, London.

Sakai, K. (2005). Environmental Design for Concrete Structures. *Journal of Advanced Concrete Technology*, 3(1), 17–28.

Sakai, K., and Buffenbarger, J.K. (2012). Concrete Sustainability Forum IV. *Concrete International*, 34(3), 41–44.

Sakai, K., and Noguchi, T. (2012). *The Sustainable Use of Concrete.* CRC Press, London.

UN. (1987). *Montreal Protocol.* United Nations, New York.

UN. (1995). *Intergovernmental Panel on Climate Change (IPCC), Second Assessment.* United Nations, New York.

UN. (1997). *Kyoto Protocol.* United Nations, New York.

Chapter 12

Codes and practice

12.1 GENERAL

Based on the observed field performance of all the concrete structures described in Chapter 2, the question may be raised why all the offshore concrete structures built for the oil and gas industry in the North Sea since the early 1970s have shown such a better durability than all the important marine concrete structures built along the Norwegian coastline during the same period. For many years, the required service life for the offshore concrete structures was typically 30 years, gradually increasing up to typically 60 years, while for all the land-based marine concrete structures, the required service life was typically 60 years, gradually increasing up to typically 100 years.

For many years, steel was the traditional structural material for the offshore oil and gas industry. Therefore, when the first concept for offshore installations in the North Sea based on concrete was introduced in the late 1960s, the offshore technical community showed much skepticism toward the use of concrete as a structural material for such a harsh and hostile marine environment (Gjørv, 1996). At the same time, however, the results and recommendations from the extensive field investigations of the more than 200 concrete structures along the Norwegian coastline had just been published (Gjørv, 1968). As outlined and discussed in Chapter 2, most these concrete structures showed a high structural capacity even after 50–60 years of the combined action of severe marine exposure and heavy structural loads. The overall good condition of these marine concrete structures contributed, therefore, to convincing the offshore technical community that concrete could also be a possible and reliable structural material for offshore installations in the North Sea.

For the international operators in the North Sea, however, a structural material that typically showed corrosion problems already after a service period of less than 10 years, and at the same time was difficult to repair, was not acceptable. Therefore, in order to get acceptance for the first offshore concrete platform for the North Sea, both increased concrete quality and increased concrete cover beyond that typically specified in current concrete codes were required. Since much of the observed durability problems for the land-based marine concrete structures also could be related to an absence of proper quality control and problems during concrete construction, very strict programs for quality control and quality assurance during concrete construction were also required before concrete could be accepted as a reliable structural material for offshore installations in the North Sea.

In order to better understand the different durability and field performance of all the offshore concrete structures in the North Sea compared to that of all the land-based marine concrete structures built along the Norwegian coastline during the same period, a brief outline and discussion of the applied codes and practice are given below.

Based on current experience with all the procedures for durability design and concrete quality control, as outlined and discussed in the previous chapters, some new recommended job specifications are also briefly outlined in the following.

12.2 CODES AND PRACTICE

12.2.1 Offshore concrete structures

In order to meet the new challenge from the more demanding offshore industry in the early 1970s, the international organization for prestressed concrete structures, Fédération Internationale de la Précontrainte (FIP), soon established a Concrete Sea Structures Commission, from which the first edition of *Recommendations for the Design and Construction of Concrete Sea Structures* was published in 1973 (FIP, 1973). The durability requirements in these new recommendations were primarily based on the experience from the extensive field investigations of all the concrete structures along the Norwegian coastline carried out during the 1960s and the recommendations from this project, but also other international experience on concrete structures in marine environments was carefully reviewed. Shortly after the FIP recommendations were published, both the Norwegian Petroleum Directorate in its regulations (NPD, 1976) and Det Norske Veritas in its rules (DNV, 1976) adopted the new and stricter durability requirements for fixed offshore concrete structures recommended by FIP.

The new durability requirements in the FIP recommendations were slightly different for the different zones of exposure, such as the submerged, splash, and atmospheric zones, respectively. For the most severe exposure in the splash zone, however, the water/cement ratio should not exceed 0.45, and preferably be 0.40 or less, subject to the attainment of adequate workability. A minimum cement content of 400 kg/m^3 should also be applied, and ordinary reinforcement and prestressing tendons should be protected by a nominal concrete cover of 75 and 100 mm, respectively.

After the first breakthrough for use of concrete in the development of the Ekofisk Oil Field, a rapid development took place, as previously described in Chapters 1 and 2. For the first concrete platform, however, it was not possible to produce a concrete with such a low water/cement ratio as 0.40 with the combined requirements of high compressive strength and 4–6% entrained air for attainment of proper frost resistance; efficient water-reducing admixtures for concrete production were not available at that time. Therefore, for the Ekofisk Tank, a concrete with a water/cement ratio of 0.45 was produced. After this structure was installed in 1973, however, rapid development on high-strength concrete took place (Gjørv, 2008), and the requirement for concrete quality successively increased from project to project. Thus, for all the further offshore concrete platforms in the North Sea, concrete with a water/cement ratio varying from 0.35 to 0.40 was produced.

Already for the Brent B Platform, which was installed in 1975, concrete with a water/cement ratio of 0.40 and a 28-day compressive strength of 48.5 MPa, in combination with 4.9% air, was produced for the splash zone, while for the concrete below water without any air entrainment, the average 28-day compressive strength was 56.9 MPa. For all the other concrete platforms produced later on, the 28-day compressive strength gradually increased up to 80 MPa for the Troll A Platform, which was installed in 1995 (Gjørv, 2008). The regular quality control of the water permeability typically showed a very dense concrete with depths of water ingress of typically less than 2 mm according to ISO/DIS 7031 (Standard Norway, 1989) and water permeability values of typically less than 10^{-13} kg/Pa·m·s (Gjørv and Løland, 1980; Gjørv, 1994).

For all the offshore concrete structures in the North Sea, very strict requirements for crack widths were also specified. Although the importance of these very strict crack requirements for the durability was not very clear, as previously discussed in Chapter 3, the consequences of these requirements were occasionally so severe that they were more decisive for the structural design and the amount of steel installed than those of the wave loads from the so-called 100-year wave.

In addition to the durability requirements, as outlined above, most of the concrete platforms built before 1980 were also protected in the splash zone by an additional solid surface coating. This coating was typically a 2–3 mm thick epoxy coating, which was continuously applied during slip forming of the structures. Due to the early application of this coating while the concrete still had an underpressure and high suction ability, a very good bond between the concrete substrate and the coating was also achieved (Chapter 5). For all the platforms, extensive programs for quality control were implemented for ensuring a best possible construction quality.

The good performance of all the offshore concrete platforms in the North Sea demonstrates that already from the early 1970s, it was possible to both design and produce very durable concrete structures even for the most aggressive and harsh marine environment. However, the above requirements for both durability and concrete quality control were based on proper utilization of current knowledge and experience. For all the professional operators in the North Sea, a high degree of safety and low operation costs of the installations were essential and of utmost importance.

Although the oldest concrete structures in the North Sea still appear to be in quite good condition, several of these structures have already gotten some extent of steel corrosion, and for some of them, very costly repairs have also been carried out (Chapter 2). In spite of both the very homogenous concrete production and the very strict programs for quality control during concrete construction, much of the corrosion problems observed later on can be ascribed to a high scatter and variability of achieved construction quality; any weaknesses in the achieved construction quality were soon revealed. It should be noted, however, that also for all the offshore concrete structures in the North Sea, the required durability was primarily based on descriptive requirements. As discussed in Chapter 6, it is not easy to verify and control any specifications based on descriptive requirements to concrete composition and execution of concrete work.

Apart from one of the concrete platforms that was built in the Netherlands with blast furnace slag cement, all the other offshore concrete platforms were also produced with pure portland cements (CEM I), which is not the best binder system for obtaining high resistance to chloride ingress (Chapter 3). Only for a few of the youngest concrete platforms was a small amount of silica fume in the concrete also applied, but this was primarily used in order to stabilize the highly flowable fresh concrete from segregation during concrete construction.

For the Brent B Platform (1975), which was not protected by any surface coating in the splash zone, a deep chloride ingress and an early stage of depassivation after approximately 20 years of exposure were observed (Chapter 2). Since the concrete for this platform was produced with a water/cement ratio of 0.40 and more than 400 kg/m^3 cement in combination with a nominal concrete cover of 75 mm, however, the durability requirements according to even the strictest European Concrete Codes were fulfilled (CEN, 2009).

12.2.2 Land-based concrete structures

For all the land-based concrete structures built along the Norwegian coastline from the early 1970s and up to the late 1980s, the requirement to concrete quality was mostly based on a given 28-day compressive strength. During the booming construction activities that took place in both Norway and many other countries during this period, a large number of both concrete harbor structures and concrete bridges were built along the Norwegian coastline. For all of these important concrete infrastructures, the requirements to compressive strength and minimum concrete cover typically varied from 30 to 40 MPa and from 25 to 50 mm, respectively. For these concrete structures, pure portland cements (CEM I) were also typically applied.

For all the above marine concrete structures built in this period, an absence of proper quality control during concrete construction in combination with poor workmanship also typically resulted in a very poorly achieved construction quality (Chapter 2). This is clearly demonstrated for the Gimsøystraumen Bridge (Figure 2.36), which was built in the northern part of Norway during 1978–1981. During the extensive repairs of this bridge, which were carried out after 11 years of service, a very deficient concrete cover was revealed, as clearly shown in Figure 2.40. In Chapter 6, the great variation of achieved concrete cover in the Gimsøystraumen Bridge was also shown in combination with similar observations in a Japanese bridge and more than 100 concrete structures in the Gulf region (Figure 6.1).

Both in Norway and many other countries, better performance of older concrete structures built before 1970 was often experienced, and this may be related to the rapid development in concrete technology that took place in the early 1970s, the consequences of which were probably not properly understood by the construction industry. Thus, until the late 1960s in Norway, a 30 MPa type of concrete was typically produced with a cement content of more than 400 kg/m^3 (Rudjord, 1967). Therefore, such a type of concrete was much more robust from a durability point of view, although the old types of concrete with a poorer concrete workability could also give poor compaction, which again could contribute to reduced durability. During the 1970s and 1980s, however, new and more finely ground portland cements were introduced, and new organic and mineral admixtures became available. At the same time, both aggregate production and concrete mixing were more optimized. Therefore, it gradually became possible to meet the specified requirement to compressive strength with lesser amounts of cement. Also, an increased production track often resulted in poorer workmanship and concrete curing. As a result, the durability properties of the concrete got successively impaired. Typically, it was not until extensive durability problems were experienced in the field before new durability specifications in several steps of new revised concrete codes were introduced. In most countries, this upgrading of the current code requirements for durability was mostly far behind the technical development and current knowledge.

Although most codes for concrete durability have been upgraded a number of times since the early 1970s, current code specifications for concrete durability are still almost exclusively based on traditional requirements to concrete composition and execution of concrete work, the results of which are neither unique nor easy to verify and control during concrete construction.

For concrete structures in marine environments, current European Concrete Codes still allow for concrete based on water/binder ratios of up to 0.45 (CEN, 2009). In the Norwegian national amendments to the European Code 206-1, however, an upper level for the water/binder ratio of 0.40 is required (Standard Norway, 2003a). It should be noted that in the early 1970s, the Norwegian Concrete Code did not have any requirement to the water/cement ratio for concrete structures in marine environments, while the requirement to minimum concrete cover was 25 mm (Standard Norway, 1973). Not until 1986 did the Norwegian Concrete Code have any limitation to the water/cement ratio for new concrete structures in marine environments, but from then on, an upper level for the water/binder ratio of 0.45 was introduced (Standard Norway, 1986).

After having experienced extensive durability problems with all their concrete bridges along the Norwegian coastline (Chapter 2), the Norwegian Public Roads Administration (NPRA) in 1988 introduced its own and stricter durability requirements for new concrete coastal bridges (NPRA, 1988). Thus, from 1988, the upper water/binder ratio was limited to 0.40, while from 1996, this upper level was reduced to 0.38 for the most exposed parts of the bridges (NPRA, 1996). In 1994, a great step forward was made when the Norwegian Public Roads Administration also introduced additional requirements in order to better ensure the specified concrete cover (NPRA, 1994). By specifying a maximum deviation of ±15 mm for the concrete cover to structural steel, an increased concrete thickness of 15 mm to the minimum concrete cover was required.

For all the concrete structures built along the Norwegian coastline from the early 1970s, it can be seen from Table 12.1 that it took a long time before similar strict durability requirements were adopted that recommended offshore concrete structures by FIP in 1973. This slow upgrading of the Norwegian Concrete Codes for durability in marine environments is also typical for most concrete codes in other countries (CEN, 2009).

Along with the rapid development of concrete technology for all the offshore structures in the North Sea, a rapid international development on high-strength concrete also took place (Gjørv, 2008). Since high-strength concrete with its low porosity and high density generally enhances the overall performance of the material, the term *high-performance concrete* (HPC) was soon introduced, which is inclusive for the term *high-strength concrete*. Internationally, therefore, *high-performance concrete* was successively specified for concrete durability rather than for concrete strength. Although a number of definitions of both *high-strength concrete* and *high-performance concrete* exist, these terms are mostly specified by a certain upper level for the water/cement ratio or water/binder ratio, but it should be noted that neither the term *water/cement ratio* nor *water/binder ratio* is unique or easy to define any longer.

Table 12.1 Development of durability requirements for concrete structures in Norwegian marine environments (splash zone)

Year	Code	Maximum water/ binder ratio	Minimum concrete cover (mm)
1973	Fédération Internationale de la Précontrainte, FIP	0.45 (0.40)	75[a]
1976	Norwegian Petroleum Directorate, NPD	0.45 (0.40)	75[a]
1976	Det Norske Veritas, DNV	0.45 (0.40)	75[a]
1973	Norwegian Concrete Code NS 3473	No requirement	25
1986	Norwegian Concrete Code NS 3420	0.45	—
1988	Norwegian Public Roads Administration, NPRA	0.40	—
1989	Norwegian Concrete Code NS 3473	—	50
1996	Norwegian Public Roads Administration, NPRA	0.38	60
2000	European Concrete Code EN-206-1	0.45	—
2003	Norwegian Concrete Code NS-EN-206-1	0.40	—
2003	Norwegian Concrete Code NS 3473	—	60

[a] Nominal concrete cover.

For many years, when concrete was mostly based on pure portland cements and simple procedures for concrete production, the concept of water/cement ratio was the fundamental basis for both characterizing and specifying concrete quality. Since a number of different cementitious materials and reactive fillers are now being applied for concrete production, the concrete properties are more and more being controlled by the various combinations of such supplementary materials. In addition, the concrete properties are also more and more being controlled by the use of various types of processed concrete aggregate, new concrete admixtures, and sophisticated production equipment.

In order to keep track of the rapid development on the use of new supplementary materials as a partial replacement for portland cement, a so-called *efficiency factor* for calculating the water/binder ratio was adopted in the European Concrete Codes, as shown in Equation 12.1:

$$W/(C + k \cdot Sm) \tag{12.1}$$

where W is water, C is portland cement, and k is an *efficiency factor* for the given supplementary material, Sm; for each type of supplementary material, a special *efficiency factor* is given.

Such an efficiency factor was originally introduced in a comprehensive research program on silica fume in concrete carried out in the late 1970s (Løland, 1981; Løland and Gjørv, 1981, 1982). In this research program, efficiency factors were introduced in order to quantify the effect of a partial replacement of the portland cement by silica fume on the various properties of the concrete. In this research program it was shown, however, that the efficiency factors varied within wide limits from one concrete property to another. Even for the same concrete property, a wide range of efficiency factors were observed, depending on the level of concrete quality. While the efficiency factors for compressive strength could typically vary from two to four, the efficiency factors for permeability and durability properties could be up to 10 times higher. The significant effect of silica fume on the durability of concrete was discussed in Chapter 3 (Figure 3.18). Therefore, silica fume is primarily used in order to enhance the durability properties of the concrete. When the concrete codes later only adopted an efficiency factor of two as a basis for the use of silica fume in concrete, the rationale for such an efficiency factor may be questioned.

On the basis of the rapid development of concrete technology in recent years, as briefly outlined above, it may be stated that the old and very simple terms *water/cement ratio* and *water/binder ratio* for characterizing and specifying concrete quality have successively lost their meaning. As a consequence, there is a great need for performance-based definitions and specifications for concrete quality. In particular, this is true for characterizing and specifying concrete durability.

Since performance-based specifications are becoming increasingly more important for infrastructure projects, the National Ready Mixed Concrete Association in the United States has already committed itself to making such specifications with its Prescription to Performance (P2P) Initiative. Such specifications will reward the producers for quality and discourage them from supplying deficient concrete (Bognacki et al., 2010).

In the United States, the rapid chloride permeability test (ASTM, 2005) was already introduced in the early 1980s by Whiting (1981), and gradually this test method was widely adopted also internationally. However, since this test method only gives an empirical *coulomb value* based on the measurement of total electrical charge passed through the concrete specimen over a short period of time, it does not give any basic information about the resistance of the concrete to chloride ingress, which can be used as an input parameter for a more proper durability analysis. In spite of this, however, it was a great step forward when this test method was introduced for specification, control, and acceptance for concrete durability.

In order to stimulate the use of high-performance concrete (HPC) for highway applications in the United States, the Federal Highway Administration, in the early 1990s, defined high-performance concrete by the following four durability and four strength parameters, which included (Goodspeed et al., 1996):

Durability properties:
- Freeze/thaw durability
- Scaling resistance
- Abrasion resistance
- Chloride permeability

Mechanical properties:
- Compressive strength
- Elasticity
- Shrinkage
- Creep

Based on requirements to each of the above parameters, four different performance grades were defined, and details of the test methods for determining the performance grades given. On this basis, applications of the various HPC grades for various exposure conditions were recommended.

In the traditional design and production of important concrete infrastructure, it appears that the importance of durability has been underestimated for many years. Therefore, public agencies are spending significant and rapidly increasing proportions of their construction budgets on repairs and maintenance of existing concrete infrastructure (Chapter 2). Enhanced durability and service life of new concrete infrastructure not only are important from an economical point of view, but also directly affect sustainability (Chapter 11).

As clearly demonstrated in Chapter 2, the achieved construction quality of new concrete structures typically shows a high scatter and variability, and in severe environments, any weaknesses and deficiencies will soon be revealed whatever durability specifications and materials have been applied. To a certain extent, a probability approach to the durability design can take the high scatter and variability of the achieved construction quality into account. However, numerical solutions alone are not sufficient to ensure the durability and service life of concrete structures in severe environments.

As shown in Chapter 4, a probability-based durability design can be used to compare and select one of several technical solutions in order to obtain a best possible durability of the given concrete structure in the given environment during the required service period. As a result, some performance-based durability requirements are established that provide the basis for concrete quality control and quality assurance during concrete construction. A final documentation of achieved construction quality and compliance with the specified durability should be key to any rational approach to more controlled and increased durability.

For all the concrete structures where the above procedures for both probability-based durability design and performance-based concrete quality control were applied (Chapter 9), the specified durability was achieved with a proper margin. For the owners of these structures, it was very important to receive documentation of the achieved construction quality and compliance with the specified durability before the structures were formally handed over from the contractors. The required documentation of achieved construction quality also clarified the responsibility of the contractors for the quality of the construction process. The required documentation of achieved construction quality clearly resulted in improved workmanship and reduced scatter and variability of the achieved construction quality.

Even if the strongest durability requirements have been specified and achieved during concrete construction, extensive experience demonstrates that for all concrete structures in chloride-containing environments, a certain rate of chloride ingress will always take place during operation of the structures. Upon completion of the above structures, therefore, it was also very important for the owners to receive a service manual for future condition assessment and preventive maintenance of the structures. It is such a service manual that helps provide the ultimate basis for achieving a more controlled durability and service life of the concrete structures.

For a more complete durability design of important concrete structures in severe environments, potential durability problems other than chloride-induced corrosion must also be properly considered and taken into account. The same is true for control of early-age cracking. For all new concrete structures, all durability requirements in current concrete codes must always be fulfilled. However, for important concrete infrastructures where high performance and safety are of special importance, efforts should be made to reach an increased and more controlled durability beyond what is possible based on current concrete codes and practice. As a basis for this, some new recommended job specifications are briefly outlined in the following.

12.3 NEW RECOMMENDED JOB SPECIFICATIONS

12.3.1 Service period

As an overall durability requirement for the given concrete structure in the given environment, a certain service period shall be required before a probability of 10% for steel corrosion is reached.

For all major concrete infrastructures, a *service period* of at least 100 years should be required before the probability of corrosion exceeds a serviceability limit state of 10% (Chapter 4). For *service periods* of more than 100 years, however, the calculation of corrosion probability gradually becomes less reliable. For *service periods* of up to 150 years, therefore, the corrosion probability should be kept as low as possible, not exceeding 10%, but in addition, some further protective measures, such as partial use of stainless steel, should also preferably be required. For *service periods* of more than 150 years, however, any calculations of corrosion probability are no longer considered valid. In order to further increase and ensure the durability, the corrosion probability should still be kept as low as possible, not exceeding 10% for a 150-year *service period*, but in addition, some additional strategies and protective measures should always be required (Chapter 5).

In marine environments, there may occasionally also be a risk for early-age exposure during concrete construction before the concrete has gained sufficient maturity and density. Also for such a case, some special precautions or protective measures should be considered. Since all additional protective measures may have implications both for the economy of the project and for the future operation of the structure, such measures should always be discussed with the owner of the structure before the special strategy and protective measure are selected.

It should be noted that a durability design based on calculations of corrosion probability does not provide any basis for prediction or assessment of service life of the given structure. Beyond onset of corrosion, a very complex deteriorating process starts with many further critical stages before the service life is reached. As soon as the first chlorides have reached embedded steel and corrosion starts, however, the owner has a problem. At an early stage of visual damage, this only represents a maintenance and cost problem, but later on, it may gradually develop into a more difficult controllable safety problem. It is in the early stage of the deteriorating process, before any chlorides have reached embedded steel and corrosion starts, that it is both technically easier and much cheaper to take necessary precautions and select proper protective measures for control of the further deteriorating process. Such control has also shown to be a much better strategy from a sustainability point of view (Chapter 11).

It should further be noted that the obtained *service periods* with a probability of corrosion of less than 10% should not be considered as real service periods for the given structures. The above *service periods* should rather be considered the result of an engineering assessment and judgment of the most important parameters related to the durability of the given structure, including their scatter and variability. In this way, a proper basis is obtained for comparing and selecting one of several technical solutions for a best possible durability of the given concrete structure in the given environment during the required *service period*. As a result, performance-based durability requirements are established, which provide the basis for performance-based concrete quality control and quality assurance during concrete construction. Also, documentation of achieved construction quality and compliance with the specified durability can be provided.

12.3.2 Achieved construction quality

Before the structure is formally handed over from the contractor to the owner, a documentation of achieved construction quality shall be required. In addition to compliance with the specified durability, this documentation shall also include a documentation of achieved in situ quality during the construction period and potential quality of the structure.

Even before the concrete is placed in the formwork, the quality of the concrete may show a high scatter and variability. Depending on a number of factors during concrete construction, the achieved quality of the finely placed concrete may show an even higher scatter and variability.

One of the most common quality problems in concrete construction, however, is the failure to meet the specified cover thickness to embedded steel. Although the specified concrete cover is normally carefully checked prior to the placing of the concrete, significant deviations can occur during concrete construction, and this does occur. The loads imposed during concrete placing may cause movement of the reinforcement in the formwork, or the chairs may have been insufficiently or wrongly placed.

Since any weaknesses and deficiencies will soon be revealed in a severe environment whatever durability specifications and materials have been applied, performance-based concrete quality control and quality assurance should be required. Current experience shows that required documentation of achieved construction quality distinctly clarifies the responsibility of the contractor for the quality of the construction process; required documentation of achieved construction quality clearly results in improved workmanship with less scatter and variability of achieved construction quality.

For the owner of the given structure, a final documentation of achieved construction quality and compliance with the specified durability should be very important, since it may have implications for both the future operation and the expected service life of the structure.

12.3.3 Condition assessment and preventive maintenance

Upon completion of the project, a service manual for regular control of the real chloride ingress during operation of the structure and recommendations for how to control this chloride ingress shall be required.

Even if the strongest durability requirements are both specified and achieved during concrete construction, extensive experience demonstrates that for all concrete structures in chloride-containing environments, a certain rate of chloride ingress will always take place during operation of the structure. As part of the durability design, therefore, a service manual for regular control of the real chloride ingress taking place during operation of the structure should be produced.

It is such a service manual that helps provide the ultimate basis for obtaining a more controlled and increased service life of the given concrete structure in the given environment. For each new condition assessment of the structure, new estimates for the probability of corrosion are developed using input parameters based on data from the observed chloride ingress. Before this probability of corrosion becomes too high, appropriate protective measures should be implemented. In this way, the need for technically difficult and very costly repairs can be reduced.

REFERENCES

ASTM. (2005). *ASTM C 1202-05: Standard Test Method for Electrical Indication of Concrete's Ability to Resist Chloride Ion Penetration.* ASTM International, West Conshohocken, PA.

Bognacki, C.J., Pirozzi, M., Marsano, J., and Baumann, W.C. (2010). Rapid Chloride Permeability Testing's Suitability for Use in Performance-Based Specifications— Concerns about Variability Can Be Mitigated. *Concrete International,* 35(5), 47–52.

CEN. (2009). *Survey of National Requirements Used in Conjunction with EN 206-1:2000,* Technical Report CEN/TR 15868. CEN, Brussels.

DNV. (1976). *Rules for the Design, Construction and Inspection of Fixed Offshore Structures.* Det Norske Veritas—DNV, Oslo.

FIP. (1973). *Recommendations for the Design and Construction of Concrete Sea Structures.* Féderation Internationale de la Précontrainte—FIP, London.

Gjørv, O.E. (1968). *Durability of Reinforced Concrete Wharves in Norwegian Harbours.* Ingeniørforlaget, Oslo.

Gjørv, O.E. (1994). Important Test Methods for Evaluation of Reinforced Concrete Durability. In *Proceedings, V. Mohan Malhotra Symposium on Concrete Technology, Past, Present and Future,* ACI SP-144, ed. P.K. Mehta, pp. 545–574.

Gjørv, O.E. (1996). Performance and Serviceability of Concrete Structures in the Marine Environment. In *Proceedings, Odd E. Gjørv Symposium on Concrete for Marine Structures,* ed. P.K. Mehta. CANMET/ACI, Ottawa, pp. 259–279.

Gjørv, O.E. (2008). High Strength Concrete. In *Developments in the Formulation and Reinforcement of Concrete,* ed. S. Mindess. Woodhead Publishing, Cambridge, UK, pp. 79–97.

Gjørv, O.E., and Løland, K.E. (1980). Effect of Air on the Hydraulic Conductivity of Concrete. In *Proceedings, First International Conference on Durability of Building Materials and Components,* ASTM STP 691, ed. P.J. Sereda and G.G. Litvan. ASTM, Philadelphia, pp. 410–422.

Goodspeed, C.H., Vanikar, S., and Cook, R.A. (1996). High Performance Concrete Defined for Highway Structures. *Concrete International,* 18(2), 62–67.

Løland, K. (1981). Mathematical Modelling of Deformational and Fracture Properties of Concrete Based on Damage Mechanical Principles—Application on Concrete with and without Addition of Silica Fume, Ph.D. Thesis. Department of Building Materials, Norwegian Institute of Technology, Trondheim.

Løland, K., and Gjørv, O.E. (1981). Silica in Concrete. *Nordisk Betong*, 25(6), 29–30 (in Norwegian).

Løland, K., and Gjørv, O.E. (1982). Condensed Silica Fume in Concrete. In *Proceedings, Nordic Research Seminar on Condensed Silica Fume in Concrete*, BML 82.610, ed. K. Løland and O.E. Gjørv. Department of Building Materials, Norwegian Institute of Technology, Trondheim, pp. 165–188.

NPD. (1976). *Regulations for the Structural Design of Fixed Structures on Norwegian Continental Shelf*. Norwegian Petroleum Directorate—NPD, Stavanger.

NPRA. (1988). *Code of Process*. Norwegian Public Roads Administration—NPRA, Oslo (in Norwegian).

NPRA. (1994). *Securing of Concrete Cover for Reinforcement*, Report 1731. Norwegian Public Roads Administration—NPRA, Oslo (in Norwegian).

NPRA. (1996). *Handbook 185*. Norwegian Public Roads Administration—NPRA, Oslo (in Norwegian).

Rudjord, A. (1967). On Aggregates for Concrete in Norway. Quality Requirements and Tests. *Nordisk Betong*, 11(3), 299–322 (in Norwegian).

Standard Norway. (1973). *NS 3473: Concrete Structures—Design and Detailing Rules*. Standard Norway, Oslo (in Norwegian).

Standard Norway. (1986). *NS 3420: Specification Texts for Buildings and Construction Works*. Standard Norway, Oslo (in Norwegian).

Standard Norway. (1989). *NS 3473: Concrete Structures—Design and Detailing Rules*. Standard Norway, Oslo (in Norwegian).

Standard Norway. (2003a). *NS-EN 206-1: Concrete—Part 1: Specification, Performance, Production and Conformity*, Amendment prA1:2003 Incorporated. Standard Norway, Oslo (in Norwegian).

Whiting, D. (1981). *Rapid Determination of the Chloride Permeability of Concrete*, Report FHWA/RD-81/119. Portland Cement Association, Skokie, IL.

Index